INTERNATIONAL CENTRE FOR MECHANICAL SCIENCES

COURSES AND LECTURES - No. 23

CEMAL ERINGEN
UNIVERSITY OF PRINCETON

FOUNDATIONS OF MICROPOLAR THERMOELASTICITY

COURSE HELD AT THE DEPARTMENT
FOR MECHANICS OF DEFORMABLE BODIES
JULY 1970

UDINE 1970

SPRINGER-VERLAG WIEN GMBH

ISBN 978-3-211-81142-9 ISBN 978-3-7091-2904-3 (eBook)
DOI 10.1007/978-3-7091-2904-3

Originally published by Springer Vienna in 1970

INTRODUCTION.

The present work is devoted to the foundation of micropolar thermoelasticity. Essentially, it is intended for the development of the exact non-linear theory. However the linear theory is produced as an approximation to the complete nonlinear theory.

Micropolar continuum mechanics is a scientific discipline concerned with the mechanics of oriented bodies whose primitive elements (roughly speaking) consist of rigid particles. Contrasted to classical continuum mechanics, the material points of a micropolar continuum are endowed with intrinsic orientations and rotary inertia. These additional degrees of freedom are believed to provide the proper physical mechanism to discuss and predict certain phenomena inherently due to the granular and molecular nature of materials. In the hierarchy of micromorphic and other nonlocal continuum theories, micropolar mechanics is a sensible first step with its solid mathematical and physical foundations and yet it is simple enough to permit serious mathematical work for the treatment of nontrivial boundary and initial value problems (in the linear theory). Applications of this theory are found and more are expected as the fields of composite materials, liquid crystals, granular solids etc. grow.

While the ideas of an oriented continu-

um can be traced all the way back to Bernoulli and Euler, in the 18th century, in connection with their work on beam theories, to MacCullogh [1839] , in connection with his theory of optics, to Duhem [1893] in thermodynamics, to Voigt [1887] in his work on crystallography and to others at the end of the nineteenth century, the first systematic work on elastic solids, bars and plates was published by E. and F. Cosserat. The important monograph of the Cosserats' [1909] was buried in the literature nearly half a century until the topic was reopened and/or rediscovered recently. Since already several expositions exist on the historical development of this field and more general theories of polar continua, cf. Eringen [1967a] , [1968] , Stojanović [1969] , we do not intend to trace the history here. However a remark may be in order : It can be said that one of the greatest contributions of the Cosserats was the introduction of a new explicit kinematics for a continuum with rigid directors which is amenable to a simple but elegant interpretation of the motion. However, the Cosserats' work possesses two basic disadvantages :

(a) It is solely based on a variational principle and thus the resultant constitutive equations are applicable to elasticity only.

(b) It lacks an explicit form for the spin energy, further the spin density defined by the hamiltonian is sufficiently vague to be of any utility.

The micropolar theory treated here

relies heavily on the papers of Eringen and his co-workers. The balance laws were already given by Eringen [1962] and the complete theory was published as a special case of a general theory of microelasticity in two papers by Eringen and Suhubi 1964a, section 6. The literature on the subject now exceeds some 1000 titles to which unfortunately no proper reference is possible here. With apologies to many distinguished authors from many countries whose important work cannot be cited properly here, I would like to proceed to develop the nonlinear theory within the context of the general theory of micromorphic materials. In the present article, I limit myself mostly to the nonlinear micropolar theory which permits an exact postulational approach. In this regard, the recent work of Kafadar and Eringen [1970] is cited here since the reader of this work will also find the relativistic theory for polar media (which may be fundamental to electromagnetic interactions).*

Chapter 1 deals with the deformation and motion of a micropolar continuum. The deformation, strain and rotation measures are introduced and their geometrical meanings are discussed. The rate tensors are calculated and compatibility conditions are obtained. Chapter 2 is devoted to the balance laws fundamental to the theo-

---

*) Eringen later [1966 b] recapitulated and introduced the present nomenclature.

ry. Conservation of mass, microinertia, balance of momenta, conservation of energy and the principle of entropy in global and local forms are given. The content of Chapter 1 and 2 are valid for all micropolar materials of any type, whether fluid, solid, viscoelastic, etc. In Chapter 3 we first present a general theory of constitutive equations and then present the constitutive equations for nonlinear thermoelastic solids. The linear theory is obtained as a special case and restricted by the condition of local material stability (nonnegative strain energy). In Chapter 4 we collect in one place the basic equations of both the nonlinear and the linear theory and obtain the field equations. The uniqueness theorem for linear thermoelastic solids is proved.

Clearly within the scope of these eight lectures no coverage is possible for the many interesting and important applications of the theory already filling a large section of the published research. Indeed this task is to be shared by other distinguished research workers on this program.

It is a great pleasure to participate in this timely program at the International Center of Mechanical Sciences at Udine. For this opportunity my deepest appreciation goes to Professors Sobrero and Olszak for their kind invitation.

I gratefully acknowledge the proof reading by Dr. C.B. Kafadar.

May, 1970 *A.C. Eringen. (Princeton Univ.)*

Chapter 1
DEFORMATION AND MOTION

1.1. Scope of the Chapter

Based on physical considerations, the mathematical model of a micropolar continuum is introduced. Coordinates, base vectors, and shifters are presented in Art. 1.2.The concept of directors, fundamental to micropolar bodies, are discussed in Art. 1.3. The motion, deformation, strains, their measures and geometrical meaning constitute the material of Arts. 1.4. and 1.5. The rate tensors appropriate to micropolar bodies are presented in Art. 1.6. The question of compatibility necessary for the single-valuedness of the deformation and rotation fields closes the chapter. These ideas are essential also to dislocations and disclinations within the context of dislocation theory.

1.2. Coordinates, Base Vectors, Shifters

The material points $\{P\}$ of a body constitute the elements of a set (called the material body) $\mathcal{B}$. These elements are considered to be known a priori from certain physical considerations that are fundamental to the structure of the mathematical theory of the physical phenomena to be studied. The set $\mathcal{B}$ is considered to be a subset of the universal set $\mathcal{U}$. This is the frame of reference or the universe for the discussion of $\mathcal{B}$. The complement of $\mathcal{B}$, denoted by $\mathcal{B}'$, is the set of all elements which are not in $\mathcal{B}$. This may be envisioned as the space surrounding the body which may contain other bodies as well, Fig. 1.2.1. Both $\mathcal{B}$ and $\mathcal{B}'$ may contain subsets. In the sequel we shall introduce some coherence (geometrical structure) to the elements of these sets so that these sets can be organized to a space. The evolution of the topological structure of the body and the interrelations of the body with the complement set $\mathcal{B}'$ is the subject of any physical theory. Thus, depending on the class of physical phenomena we intend to study, we need to establish :

(i) The physical properties of the elements of body $\mathcal{B}$ and the rules of operations (the physical laws) that they are subject to

(ii) The topological structure of $\mathcal{B}$

(iii) The model of interactions among the elements of $\mathcal{B}$ with the universe (or with $\mathcal{B}'$).

Here we are concerned with the mechanics of certain special types of deformable bodies, namely, micropolar bodies. As we shall see, this requires that we attribute to an element of $\mathcal{B}$ two independent physical notions - the mass and the inertia. The laws of motion for mass and inertia (to be postulated in the following section) provide item (i) above. The topological structure of the body will be given by its geometrical structure. To this end we shall introduce a metric space. Finally the model of interaction of the body with its exterior will be made by the consideration of the equivalent physical and geometrical effects (e.g., body and surface loads, constraints, etc.).

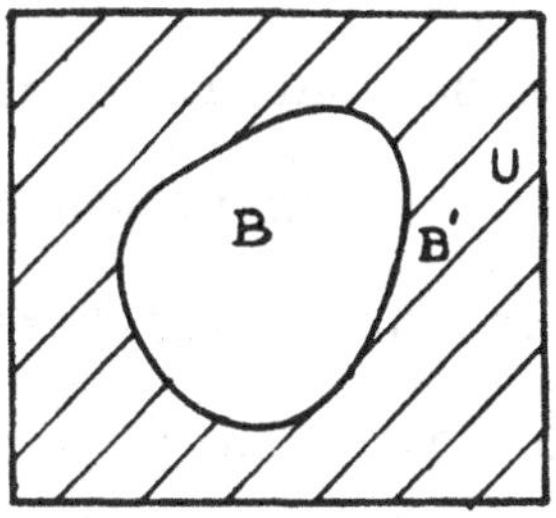

Fig. 1.2.1

Complement of B.

According to this program, first we establish a one-to-one correspondence between each point P of the body B at time $t=0$, and a set E^3 of all sets of triples $(X^1, X^2, X^3,)$ where X^1, X^2 and X^3 or simply $X^K (K=1,2,3)$ are real numbers. The set E^3 may be endowed with various mathematical structures. For example, it may be a metric space of some kind. In fact we consider E^3 to be a coordinate manifold in a Euclidean three space. When we wish to distinguish the curvilinear coordinates from rectangular coordinates, in place of X^K we write Z^K or in explicit form X, Y, Z for Z^1, Z^2 and Z^3 respectively. This picture is identical to the one used in classical continuum mechanics. The passage from rectangular coordinates to curvilinear coordinates is made through a mapping which is one-to-one

$$Z^K = Z^K(X^1, X^2, X^3) \Leftrightarrow X^K = X^K(Z^1, Z^2, Z^3) \qquad (1.2.1)$$

so that the jacobian *)

$$J \equiv \det(\partial Z^K / \partial X^L) \neq 0 \qquad (1.2.2)$$

*) For simplicity, we use the same symbol for the mapping and its value. When this point requires special emphasis, we shall change the kernel letter for the function.

in $\mathcal{B}$ at all points.

The mapping (1.2.1) is clearly assumed to possess continuous partial derivatives. In continuum mechanics such continuity and differentiability requirements are made so often that it is tiresome to repeat these obvious situations each time. Except when it is especially important to emphasize this point, the reader is burdened with the task of understanding that all partial derivatives needed in any expression are assumed to be continuous unless the contrary is mentioned.

The mapping (1.2.1) is equivalent to a graph of networks at each point with three surfaces $X^1 =$ constant, $X^2 =$ = constant, and $X^3 =$ constant intersecting at $\mathcal{P}$ which produce three curvilinear lines, Fig. 1.2.2.

With the identification of the material points of $\mathcal{B}$ with E^3, the body is made into a subset of a Euclidean metric space available for our study. Thus for example, three independent unit vectors, $\underset{\sim}{I}_K$, exist such that the material point $\mathcal{P}$ may be located by a vector (called the position vector)

$$\underset{\sim}{P} = Z^K \underset{\sim}{I}_K = Z^1 \underset{\sim}{I}_1 + Z^2 \underset{\sim}{I}_2 + Z^3 \underset{\sim}{I}_3 \tag{1.2.3}$$

spanned from the origin O of the rectangular coordinates to P. In (1.2.3) and throughout these lectures, we shall employ the summation convention over diagonally repeated indices. When a suspension of summation is desired, we shall underscore the repeated indices. If the same index appears more than twice, to avoid the ambiguity we shall revert to the summation sign. Thus, for example,

$$Z^{\underline{K}}\underset{\sim}{I}_{\underline{K}} \equiv \text{ any one of } \left(Z^1\underset{\sim}{I}_1, Z^2\underset{\sim}{I}_2, Z^3\underset{\sim}{I}_3\right)$$

however, $A^K{}_K Z^K$ is ambiguous but $\sum_{K=1}^{3} A^K{}_K Z^K$ is not.

Three noncoplanar base vectors, $\underset{\sim}{G}_K$, defined by (Fig. 1.2.2)

$$\underset{\sim}{G}_K = \frac{\partial \underset{\sim}{P}}{\partial X^K} = \frac{\partial Z^L}{\partial X^K}\underset{\sim}{I}_L = Z^L{}_{,K}\underset{\sim}{I}_L \tag{1.2.4}$$

are tangent to the curvilinear coordinate lines at P. Here and throughout, an index followed by a comma represents partial differentiation.

The metric tensor G_{KL} is constructed by forming the scalar product

$$G_{KL} \equiv \underset{\sim}{G}_K \cdot \underset{\sim}{G}_L \tag{1.2.5}$$

which is fundamental in calculating length and angles. Thus, for example, the square of the element of arc is given by

$$ds^2 = d\underset{\sim}{P} \cdot d\underset{\sim}{P} = \frac{\partial \underset{\sim}{P}}{\partial X^K} \cdot \frac{\partial \underset{\sim}{P}}{\partial X^L}\, dX^K dX^L = G_{KL}\, dX^K dX^L . \tag{1.2.6}$$

By use of (1.2.4) we see that

$$G_{KL} = \delta_{MN} Z^M{}_{,K} Z^N{}_{,L} \tag{1.2.7}$$

where $\delta_{MN} = \underset{\sim}{I}_M \cdot \underset{\sim}{I}_N$ is the Kronecker delta which is equal to 1 when $M = N$ and zero otherwise.

Similarly the angle between two base vectors is given by

(1.2.8) $$\cos(\underset{\sim}{G}_K \cdot \underset{\sim}{G}_L) = \frac{\underset{\sim}{G}_K \cdot \underset{\sim}{G}_L}{|\underset{\sim}{G}_{\underline{K}}|\ |\underset{\sim}{G}_{\underline{L}}|} = \frac{G_{KL}}{\sqrt{G_{\underline{KK}} G_{\underline{LL}}}}$$

From this it is clear that the curvilinear coordinates X^K are orthogonal if and only if $G_{KL} = 0$, for $K \neq L$. There is no particular difficulty in calculating the angle between any two directions at P.

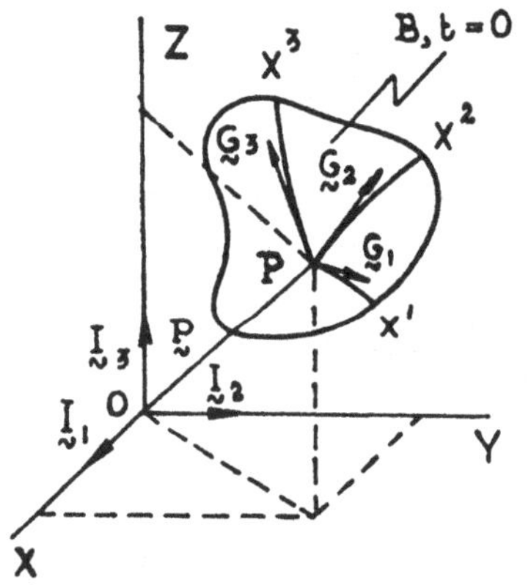

Fig. 1.2.2. Curvilinear Coordinates.

The contravariant components, G^{KL}, of the metric tensor are determined from the reduced cofactor of the matrix $\|G_{KL}\|$, i.e.,

$$G^{KL} \equiv \frac{\text{cofactor of } G_{KL}}{\det G_{KL}} \quad (1.2.9)$$

which satisfy the nine equations

$$G^{KM} G_{ML} = \delta^K{}_L \quad (1.2.10)$$

The <u>reciprocal base vectors</u> $\underset{\sim}{G}^K$ are obtained from

(1.2.11) $$\underset{\sim}{G}^K = G^{KL} \underset{\sim}{G}_L \quad .$$

They are mutually perpendicular to the set $\underset{\sim}{G}_K$, since

(1.2.12) $$G^K \cdot \underset{\sim}{G}_L = \delta^K{}_L \quad .$$

From (1.2.11) we also see that

$$G^{KL} = \underset{\sim}{G}^{K} \cdot \underset{\sim}{G}^{L} . \tag{1.2.13}$$

By use of the metric tensors G_{KL} and G^{KL} we can raise and lower indices of vectors and tensors associated with $\mathbf{B}$ .For example, a vector $\underset{\sim}{V}$ at P may be expressed in terms of its <u>contravariant</u> components, V^{K} and <u>covariant</u> components, V_{K}, at P by

$$\underset{\sim}{V} = V^{L} \underset{\sim}{G}_{L} = V_{L} \underset{\sim}{G}^{L} . \tag{1.2.14}$$

By taking the scalar product of this with $\underset{\sim}{G}^{K}$ and $\underset{\sim}{G}_{K}$ we obtain

$$V^{K} = \underset{\sim}{V} \cdot \underset{\sim}{G}^{K} = G^{KL} V_{L} \qquad V_{K} = \underset{\sim}{V} \cdot \underset{\sim}{G}_{K} = G_{KL} V^{L} . \tag{1.2.15}$$

All these are well-known from tensor algebra.

The points of B may be referred to another set of curvilinear coordinates x^{k}, (k = 1,2,3). If in this system the base vectors are $\underset{\sim}{g}_{k}$ and their reciprocals are $\underset{\sim}{g}^{k}$, we may express $\underset{\sim}{G}_{K}$ and $\underset{\sim}{G}^{K}$ in terms of these new base vectors in a unique way, i.e.,

$$\underset{\sim}{G}_{K}(\underset{\sim}{X}) = g_{K}{}^{k}(\underset{\sim}{X}, \underset{\sim}{x}) \underset{\sim}{g}_{k}(\underset{\sim}{x}) , \quad \underset{\sim}{g}_{k}(\underset{\sim}{x}) = g^{K}{}_{k}(\underset{\sim}{X}, \underset{\sim}{x}) \underset{\sim}{G}_{K}(\underset{\sim}{X}) \tag{1.2.16}$$

where

$$g_{K}{}^{k} = \underset{\sim}{G}_{K} \cdot \underset{\sim}{g}^{k} = g^{kl} G_{KL} g^{L}{}_{l} . \tag{1.2.17}$$

The two-point tensors $g^{K}{}_{k}$, $g_{K}{}^{k}$ (and g^{Kk} and g_{Kk} which may be defined by raising and lowering the indices k and K by use of g^{kl}, g_{kl} and G^{KL}, G_{KL} are called the shifters. By means of these tensors, we shift a vector in a parallel manner from a point P to another point p .

This point is clarified especially when both coordinates are rectangular. In this case $g_{kK} = \underset{\sim}{i}_k \cdot \underset{\sim}{I}_K$ where $\underset{\sim}{i}_k$ are the new cartesian unit vectors associated with the rectangular coordinates z^k. In these rectangular coordinates, for a vector $\underset{\sim}{V}$ at $\underset{\sim}{P}$ and $\underset{\sim}{v}$ at $\underset{\sim}{p}$ we have, Fig. 1.2.3.

$$\underset{\sim}{V} = V^L \underset{\sim}{I}_L , \quad \underset{\sim}{v} = v^l \underset{\sim}{i}_l$$

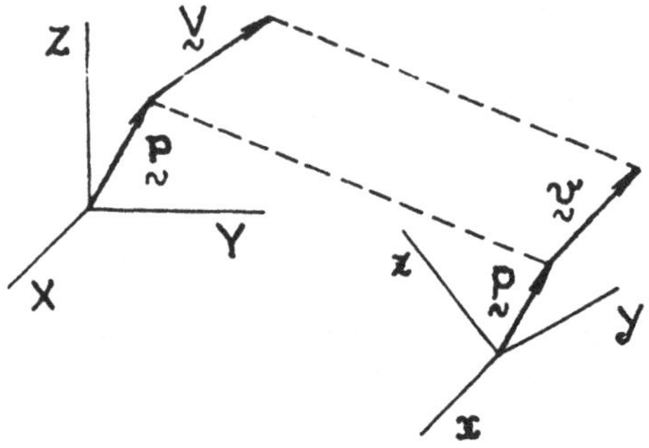

Fig. 1.2.3. Shifters.

If $\underset{\sim}{V} = \underset{\sim}{v}$ we see that

$$V^L \underset{\sim}{I}_L = v^l \underset{\sim}{i}_l \quad .$$

By scalar multiplication with $\underset{\sim}{I}^K \equiv \underset{\sim}{I}_K$,

$$V^K = \delta^K{}_k v^k \quad , \quad \delta^K{}_k \equiv \underset{\sim}{I}^K \cdot \underset{\sim}{i}_k \quad . \tag{1.2.18}$$

Here $\delta^K{}_k$ is the usual Kronecker delta replacing $g^K{}_k$ if an only if the z^k and Z^K are coincident. It shifts the vector $\underset{\sim}{v}$ from the point p to the point P.

1.3. Directors

In micropolar theory it is posited that the material points of body are endowed with <u>orientation</u>. Mathematically this means that to each point $\underset{\sim}{X}$ of the body B at time $t = 0$ there is attached three rigid vectors, $\underset{\sim}{\chi}_K$, Fig. 1.3.1. These

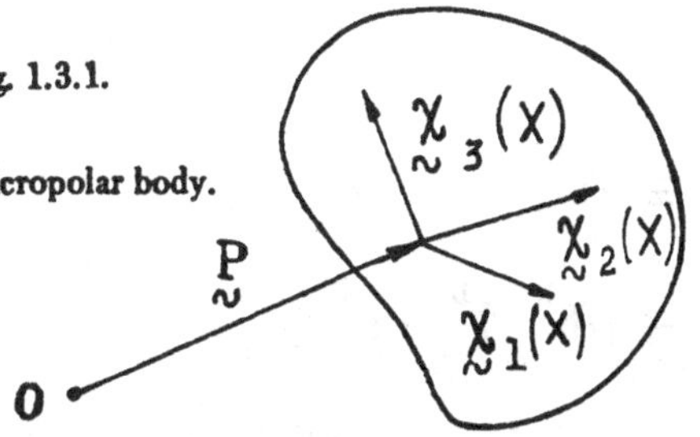

Fig. 1.3.1.

Micropolar body.

vectors change from point to point, i.e.,

$$\underset{\sim}{\chi}_K = \underset{\sim}{\chi}_K(\underset{\sim}{X}) , \quad K = 1, 2, 3. \tag{1.3.1}$$

Physically, we envision the body as consisting of small rigid bodies. In terms of its components $\chi^K{}_L$, we may write

$$\underset{\sim}{\chi}_K = \chi^L{}_K \underset{\sim}{G}_L \ . \tag{1.3.2}$$

In particular, if we wish, we can take $\underset{\sim}{\chi}_K$ at $\mathbf{P}$ parallel to $\underset{\sim}{G}_K$. The nine components $\chi^L{}_K$ of $\underset{\sim}{\chi}_K$ are not all arbitrary. In fact the rigidity of $\underset{\sim}{\chi}_K$ implies that

$$G_{KL}\chi^K{}_M\chi^L{}_N = G_{MN} \ . \tag{1.3.3}$$

Thus $\chi^K{}_L$ is an orthogonal tensor. Every proper orthogonal tensor can be expressed in the form

$$\chi_{KL} = \exp(\Phi_{KL}) \quad , \qquad \Phi_{KL} \equiv -\varepsilon_{KLM}\Phi^M \tag{1.3.4}$$

where Φ^M is an absolute vector and ε_{KLM} is the permutation tensor. Here $(1.3.4)_1$ is understood in the sense of matrix functions, i.e.,

$$\underset{\sim}{\chi} = \underset{\sim}{I} + \frac{1}{2!}\underset{\sim}{\Phi} + \frac{1}{3!}\underset{\sim}{\Phi}^2 + \cdots \tag{1.3.5}$$

For a second order tensor $\underset{\sim}{\Phi}$ we have the Cayley-Hamilton theorem

$$\underset{\sim}{\Phi}^3 - \mathrm{I}\,\underset{\sim}{\Phi}^2 + \mathrm{II}\,\underset{\sim}{\Phi} - \mathrm{III}\,\underset{\sim}{I} = \underset{\sim}{0} \tag{1.3.6}$$

where $\underset{\sim}{I}$ is the unit tensor and

$$I = tr\,\underset{\sim}{\Phi} \equiv \Phi^K{}_K$$

(1.3.7) $$II = \frac{1}{2}\left[(tr\,\underset{\sim}{\Phi})^2 - tr\,\underset{\sim}{\Phi}^2\right] = \frac{1}{2}\left(\Phi^K{}_K \Phi^L{}_L - \Phi^K{}_L \Phi^L{}_K\right)$$

$$III = \frac{1}{3}\left[tr\,\underset{\sim}{\Phi}^3 - \frac{3}{2}\,tr\,\underset{\sim}{\Phi}^2\,tr\,\underset{\sim}{\Phi} + \frac{1}{2}(tr\,\underset{\sim}{\Phi})^3\right] = det\,\Phi^K{}_L$$

are the invariants of $\underset{\sim}{\Phi}$.

By use of (1.3.6) we can express all terms $\underset{\sim}{\Phi}^N$, $N>2$ in terms of $\underset{\sim}{I}$, $\underset{\sim}{\Phi}$ and $\underset{\sim}{\Phi}^2$ with coefficients that are polynomials in the invariants of $\underset{\sim}{\Phi}$. Thus (1.3.5) can be written as

(1.3.8) $$\underset{\sim}{\chi} = \alpha_0 \underset{\sim}{I} + \alpha_1 \underset{\sim}{\Phi} + \alpha_2 \underset{\sim}{\Phi}^2$$

where α_0, α_1 and α_2 are polynomials in the invariants (1.3.7). But from (1.3.7) and $(1.3.4)_2$ we have

(1.3.9) $$I = \Phi^K{}_K = 0 \quad , \quad II = G_{MN}\Phi^M \Phi^N, \; III = 0$$

so that (1.3.6) in this case reduces to

(1.3.10) $$\underset{\sim}{\Phi}^3 = -II\,\underset{\sim}{\Phi}\,.$$

We now determine α_0, α_1 and α_2 by forming

$$\underset{\sim}{\chi}\,\underset{\sim}{\chi}^T = \underset{\sim}{I}$$

where $\underset{\sim}{\chi}^T$ is the transpose of $\underset{\sim}{\chi}$. Remembering $\underset{\sim}{\Phi}^T = -\underset{\sim}{\Phi}$ and by use of (1.3.8)

$$\underset{\sim}{I} = \alpha_0^2 \underset{\sim}{I} + \left(2\alpha_0\alpha_2 - \alpha_1^2 - II\,\alpha_2^2\right)\underset{\sim}{\Phi}^2$$

where we used (1.3.10). From this it follows that

$$\alpha_0^2 = 1 \;, \quad \alpha_1^2 = 2\alpha_0\alpha_2 - II\,\alpha_2^2 \;.$$

Upon setting det $\underset{\sim}{\chi} = 1$, we obtain one more relation which leads to the determination of α_0, α_1 and α_2 and results in

$$\chi^K{}_L = \cos\theta\,\delta^K{}_L - \sin\theta\,\varepsilon^K{}_{LM}N^M + (1-\cos\theta)N^K N_L \qquad (1.3.11)$$

where

$$\theta \equiv \sqrt{\Phi_K \Phi^K} \geq 0$$
$$(1.3.12)$$
$$N_K \equiv \Phi_K/\theta \quad .$$

The scalar θ and the vector N_K defined above possess simple geometrical interpretations : Clearly θ is the magnitude of the vector Φ_K and N_K is the unit vector along Φ_K. We therefore have

Theorem 1. Every proper orthogonal transformation can be expressed in terms of a vector, Φ_K in the form (1.3.11).

Thus to each material point of the body there is attached a vector whose magnitude and direction may change from point to point. Physically, therefore, the material points of a micropolar body are rigid particles with orientations. There are many materials of this type which occur in nature. For example, liquid crystals possess dipolar elements in the shape of short bars and cigars and layers of cigars. In particular, nematic and semectic liquid crystals, in their natural state, exhibit an ideal picture of a micropolar body consisting of cigar-like elements oriented along a common direction. Particulate composite materials consist of rigid particles or short fibers of very high strength (e.g., graphite or boron-epoxy fibers) mixed in a plastic matrix. In fact all materials can be said to consist of such a structure to an extent. The physical phenomena arising from the presence of the orientations of material points are many and they may come into play in the anisotropic properties of solids and in wave propagations at certain high frequency ranges.

1.4. The Motion, Deformation, Strain Measures

The motion of a micropolar body is characterized by the time evolution of the position and the directors of the material points. Mathematically, we express this by two sets of a one-parameter family of mappings

(1.4.1) $$x^k = x^k(\underset{\sim}{X}, t) \quad , \quad (k = 1, 2, 3)$$

(1.4.2) $$\chi^k{}_K = \chi^k{}_K(\underset{\sim}{X}, t) \quad , \quad (k, K = 1, 2, 3) \; .$$

The first one of these, (1.4.1), is the classical expression of motion, namely, that a material point $\underset{\sim}{X}$ at time t occupies the spatial position x^k. This mapping is one-to-one for all times, i.e., the inverse mapping

(1.4.3) $$X^K = X^K(\underset{\sim}{x}, t) \quad , \quad K = 1, 2, 3$$

is the unique solution of (1.4.1) for each time t. Thus the jacobian

$$J \equiv \det x^k{}_{,K} \neq 0 \quad \text{in } B \tag{1.4.4}$$

except possibly at some singular surfaces, lines and points of sets of measure zero. Thus excluding such singular submanifolds, (1.4.1) and (1.4.3) express the physical assumption of <u>continuity</u>, <u>indestructibility</u> and <u>impenetrability</u> of matter. No region of positive finite volume of B is deformed into one of zero or infinite volume. Also every region goes into a region, every surface into a surface, and every curve into a curve.

Equations (1.4.2) are the time evolution of the directors attached to the material point $\underset{\sim}{X}$. It is the mathematical expression of the rigid rotatory motion of the material particles of the body. The assumption of rigidity imposes the restriction that $\chi^k{}_K$ is orthogonal at all times. Thus the inverse matrix $\overset{-1}{\chi}{}^K{}_k$ (which is the transpose of $\chi_k{}^K$) exists such that

$$\chi^k{}_K \overset{-1}{\chi}{}^K{}_l = \delta^k{}_l \quad , \quad \overset{-1}{\chi}{}^K{}_k \chi^k{}_L = \delta^K{}_L \ . \tag{1.4.5}$$

Any one of these two sets of nine linear equations may be solved to give

$$\overset{-1}{\chi}{}^K{}_k = \frac{1}{2!} \varepsilon^{KLM} \varepsilon_{klm} \chi^l{}_L \chi^m{}_M \tag{1.4.6}$$

since we have

(1.4.7) $$\det \chi^k_{\ K} = \det \overset{-1}{\chi}{}^K_{\ k} = 1 \ .$$

If we set

(1.4.8) $$\chi^k_{\ K}(\underset{\sim}{X},0) = g^k_{\ K}$$

where $g^k_{\ K}$ are the shifters, then (1.4.5) to (1.4.7) are satisfied automatically for $t = 0$. This identification, however, is not
essential for the theory. It produces some simplifications in
the physical picture of the deformation. For convenience we also
introduce two other tensors

(1.4.9) $$\chi^k_{\ \ell} \equiv \chi^k_{\ K} g^K_{\ \ell} \quad , \quad \chi^K_{\ L} \equiv g^K_{\ k} \chi^k_{\ L} \ .$$

In the same way as in section (1.3) we can write

(1.4.10) $$\chi_{k\ell} = \exp \varphi_{k\ell} \quad , \quad \varphi_{k\ell} = -\varepsilon_{k\ell m} \varphi^m$$

and find the expression

(1.4.11) $$\chi^k_{\ \ell} = \cos \theta \, \delta^k_{\ \ell} - \frac{\sin \theta}{\theta} \varepsilon^k_{\ \ell m} \varphi^m + \frac{1-\cos\theta}{\theta^2} \varphi^k \varphi_\ell$$

where

(1.4.12) $$\theta \equiv \sqrt{\varphi^k \varphi_k} \geq 0$$

is the <u>angle of rotation</u>. Clearly when $\varphi^k = 0$, $\chi^k_{\ \ell} = \delta^k_{\ \ell}$ and the
directors are not rotated. Further, when $\chi^k_{\ \ell} = \delta^k_{\ \ell}$, $\varphi^k = 0$. Thus we
have

Theorem. The necessary and sufficient condition that the directors remain parallel to their original directions is the vanishing of φ^k.

It must be remembered, at this point, that the angle of rotation of the macrorotation (the classical rotation tensor) due to change in the line elements need not vanish at all. In fact, here we take cognizance of the difference between these two rotations : the rotation tensor of classical continuum mechanics and the microrotation tensor $\underset{\sim}{\chi}$ of micropolar theory. This point will become clear later.

In continuum mechanics, deformation gradients play a central role. These are defined by

$$x^k{}_{,K} \equiv \frac{\partial x^k}{\partial X^K} \quad , \quad X^K{}_{,k} \equiv \frac{\partial X^K}{\partial x^k} \tag{1.4.13}$$

subject to

$$x^k{}_{,K} X^K{}_{,\ell} = \delta^k{}_\ell \quad , \quad X^K{}_{,k} x^k{}_{,L} = \delta^K{}_L \ . \tag{1.4.14}$$

The solution of one of these sets of nine equations gives one of (1.4.13) and (1.4.13) in terms of the other, e.g.,

$$X^K{}_{,k} = \frac{1}{2J} \varepsilon^{KLM} \varepsilon_{k\ell m} x^\ell{}_{,L} x^m{}_{,M} \tag{1.4.15}$$

where

$$J \equiv \sqrt{\frac{g}{G}} \det(x^k{}_{,K}) = \frac{1}{6} \varepsilon^{KLM} \varepsilon_{k\ell m} x^k{}_{,K} x^\ell{}_{,L} x^m{}_{,M} \ . \tag{1.4.16}$$

By differentiation of (1.4.16) we have the classical identity of

Jacobi

(1.4.17) $$\partial j/\partial x^k_{,K} = \text{cofactor}\ x^k_{,K} = j X^K_{,k} \ .$$

The square of the arc length referred to material and spatial coordinates is given by

(1.4.18) $$ds^2 = g_{k\ell}\, dx^k dx^\ell \ , \quad dS^2 = G_{KL}\, dX^K dX^L \ .$$

The motion (1.4.1) and inverse motion (1.4.3) provide the connection of a material line element dX^K to its deformed spatial image by

(1.4.19) $$dx^k = x^k_{,K}\, dX^K \ , \quad dX^K = X^K_{,k}\, dx^k \ .$$

Substituting these into (1.4.18) we get

(1.4.20) $$ds^2 = C_{KL}\, dX^K dX^L \ , \quad dS^2 = c_{k\ell}\, dx^k dx^\ell$$

where

(1.4.21) $$C_{KL}(\underset{\sim}{X}, t) \equiv g_{k\ell}(\underset{\sim}{x})\, x^k_{,K}\, x^\ell_{,L}$$

(1.4.22) $$c_{k\ell}(\underset{\sim}{x}, t) \equiv G_{KL}(\underset{\sim}{X})\, X^K_{,k}\, X^L_{,\ell}$$

are respectively the Green and Cauchy deformation tensors of classical continuum mechanics. Two other important deformation tensors used (cf. Eringen 1962, art. 4) are the inverse of the above tensors

(1.4.23) $$\overset{-1}{C}{}^{KL} \equiv g^{k\ell}\, X^K_{,k}\, X^L_{,\ell}$$

$$\overset{-1}{C}{}^{kl} = G^{KL} x^k_{,K} x^l_{,L} \tag{1.4.24}$$

called respectively the Piola and Finger deformation tensors.

In micropolar continuum mechanics, we have two sets of kinematical degrees of freedom, the macromotion (1.4.1) and the micromotion (1.4.2). Any local measure of deformation must not only involve the macrodeformation gradients (1.4.13) but also $\chi^k{}_K$ and its gradients. Thus the following three sets of quantities are necessary for the description of the deformation of a material particle of micropolar continua :

$$x^k_{,K} \quad , \; \chi^k{}_K \, , \; \chi^k{}_{K:L} \tag{1.4.25}$$

where a colon denotes total covariant differentiation (cf. Eringen, 1962, Appendix). The necessity of $\chi^k{}_K$ arises from the fact that without it we cannot describe the orientations of the particles and that of $\chi^k{}_{K:L}$ is needed to find out how the directors of neighboring particles change. Clearly, as we do not need all nine $x^k_{,K}$ to obtain the length and angle changes in classical continuum mechanics, so we do not need all the components of $\chi^k{}_K$ and $\chi^k{}_{K:L}$ for the description of the length and angle changes in a micropolar body. Thus the question arises : what tensorial entities with a minimal number of independent components shall be used ? There exist many ways of constructing sets of entities. Just as in the case of classical strain measures, this nonunique answer constitutes no threat to the theory. Each choice may provide a special convenience. Eringen and Suhubi [1964 a] through their more general theory of microelasticity gave several forms. Eringen [1966 b] for the linear micropolar theory of viscoelasticity found some more convenient to use than others. The Cosserats [1909] indicated another set. By use of the Clausius-Duhem inequality we shall arrive at another set which was obtained recently by Kafadar and Eringen [1970]. All these findings indicate that the deformation of a micropolar continuum can be characterized completely by two sets of strain measures.

$$\mathfrak{C}_{KL} \equiv x^k{}_{,K} \chi_{kL} \tag{1.4.26}$$

$$\Gamma_{KL} \equiv \frac{1}{2} \varepsilon_{KMN} \chi^{lM}{}_{:L} \chi_l{}^N . \tag{1.4.27}$$

We shall call $\mathfrak{C}_{KL}$ the <u>Cosserat deformation tensor</u> and Γ_{KL} the <u>wryness tensor</u>.

Since it is simple, we present here briefly the method used by Eringen and Suhubi. This involves the use of the internal energy (or free energy) function $\mathcal{E}$ relevant to the theory of micropolar elasticity. In the spirit of the theory of elasticity $\mathcal{E}$ is a function of the 45 quantities (1.4.25)

$$\mathcal{E} = \mathcal{E}\left(x^k{}_{,K}, \chi^k{}_K, \chi^k{}_{K:L}\right)$$

subject to the requirement that this function must be form invariant under rigid motion, i.e.,

$$\varepsilon(x^k_{,K}, \chi^k_K, \chi^k_{K:L}) = \varepsilon(Q^k_\ell x^\ell_{,K}, Q^k_\ell \chi^\ell_K, Q^k_\ell \chi^\ell_{K:L})$$

for the proper orthogonal group of transformations Q^k_ℓ of the spatial frame of reference. This, according to a theorem of Cauchy (cf. Eringen, 1967, App. B) implies that ε is a function of the joint invariants formed from the set (1.4.25). But ε is determined not by 45 independent quantities but by the 21 quantities $x^k_{,K}$ φ_k and $\varphi_{k:K}$. Thus the minimum function basis of ε has $21 - 3 = 18$ independent members. Since the set (1.4.26) and (1.4.27) constitutes such a basis, we see that only two asymmetric tensors are needed for the description of the deformation measures of micropolar elasticity.

At this point perhaps the reader wonders why we have discarded the Green strain measure. Note that

$$C_{KL} = g_{k\ell} x^k_{,K} x^\ell_{,L} = G^{MN} \mathfrak{C}_{KM} \mathfrak{C}_{LN}$$

so that C_{KL} is not independent of $\mathfrak{C}_{KL}$.

Similarly, spatial deformation tensors can be introduced. One such set is

$$c^{k\ell} \equiv x^k_{,K} \chi^{\ell K} = \chi^{kK} \chi^{\ell L} \mathfrak{C}_{LK} \tag{1.4.28}$$

$$\gamma^{k\ell} \equiv \chi^{kK} \chi^{\ell L} \Gamma_{KL} \quad . \tag{1.4.29}$$

Similar to C_{KL} inverses of $\mathfrak{C}_{KL}$ and $c^{k\ell}$ exist and may prove to be convenient for some purposes

(1.4.30) $$\overset{-1}{\mathfrak{C}}{}^{KL} = \chi^{kK} X^{L}{}_{,k}$$

(1.4.31) $$\overset{-1}{c}_{k\ell} = \chi_{kK} X^{K}{}_{,\ell} \quad .$$

The spatial measures of deformation (1.4.28), (1.4.29), and (1.4.31) are found especially useful in dealing with isotropic materials.

If desired, the displacement vector $\underset{\sim}{u}$ may be introduced by

(1.4.32) $$\underset{\sim}{u} = \underset{\sim}{p} - \underset{\sim}{P} \quad \text{or} \quad u^{k} = p^{k} - P^{K} g^{k}{}_{K}$$

where $\underset{\sim}{p}$ and $\underset{\sim}{P}$ are the spatial and material position vectors, respectively. By use of (1.4.32) and (1.4.11) the deformation tensors can be expressed solely in terms of $\underset{\sim}{u}$ and the rotation tensor $\underset{\sim}{\varphi}$. The resulting expressions are complicated and of no particular value in the nonlinear theory. For the linear theory they reduce to simple forms

(1.4.33) $$c_{k\ell} - g_{k\ell} \equiv \varepsilon_{k\ell} \simeq u_{k:\ell} - \varepsilon_{\ell k m} \varphi^{m} \equiv \tilde{\varepsilon}_{k\ell}$$

(1.4.34) $$\gamma_{k\ell} \simeq \varphi_{k;\ell} \equiv \tilde{\gamma}_{k\ell}$$

which were employed by Eringen, [1967 b], in his theory of micropolar viscoelasticity as the fundamental tensors for the strain measures.

Finally we note that $\underset{\sim}{\mathfrak{C}}$ and $\underset{\sim}{c}$, $\underset{\sim}{\gamma}$ and $\underset{\sim}{\Gamma}$ differ by but an orthogonal transformation. It follows that their invariants are the same, e.g.,

$$\mathrm{tr}\,\underset{\sim}{\mathfrak{C}} = \mathrm{tr}\,\underset{\sim}{\mathfrak{c}} \;,\quad \mathrm{tr}\,\underset{\sim}{\mathfrak{C}}^2 = \mathrm{tr}\,\underset{\sim}{\mathfrak{c}}^2 \;,\quad \mathrm{tr}\,\underset{\sim}{\mathfrak{C}}^3 = \mathrm{tr}\,\underset{\sim}{\mathfrak{c}}^3$$

$$\mathrm{tr}\,\underset{\sim}{\mathfrak{C}}\underset{\sim}{\mathfrak{C}}^T = \mathrm{tr}\,\underset{\sim}{\mathfrak{c}}\underset{\sim}{\mathfrak{c}}^T \;,\quad \mathrm{tr}\,\underset{\sim}{\mathfrak{C}}^2\underset{\sim}{\mathfrak{C}}^T = \mathrm{tr}\,\underset{\sim}{\mathfrak{c}}^2\underset{\sim}{\mathfrak{c}}^T,\quad \mathrm{tr}\,\underset{\sim}{\mathfrak{C}}^2(\underset{\sim}{\mathfrak{C}}^T)^2 = \mathrm{tr}\,\underset{\sim}{\mathfrak{c}}^2(\underset{\sim}{\mathfrak{c}}^T)^2 \tag{1.4.35}$$

and similarly for $\underset{\sim}{\gamma}$ and $\underset{\sim}{\Gamma}$.

1.5. Geometrical Meaning of Strain Measures

The geometrical significance of the Cosserat deformation tensor may be seen as follows : A material curve $\mathcal{C}$ characterized by $X^K = X^K(\lambda)$ in **B** deforms to the spatial curve **c** , Fig. 1.5.1.

$$x^k = x^k(X^K(\lambda), t) \; .$$

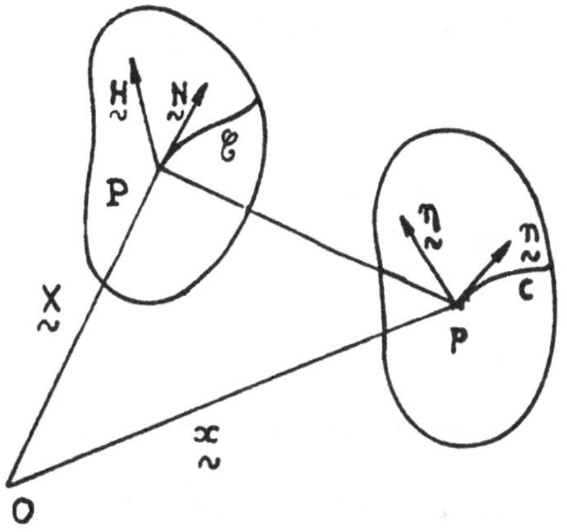

Fig. 1.5.1.

Deformation of a curve C under the macrobation.

The unit tangents to these curves are given by

$$N^K = \frac{\partial X^K}{\partial \lambda}\left(G_{RS}\frac{\partial X^R}{\partial \lambda}\frac{\partial X^S}{\partial \lambda}\right)^{-1/2}, \quad \mathcal{C}$$

$$n^k = \frac{\partial x^k}{\partial \lambda}\left(C_{RS}\frac{\partial X^R}{\partial \lambda}\frac{\partial X^S}{\partial \lambda}\right)^{-1/2}, \quad c \tag{1.5.1}$$

From these we see that

$$n^k = \Lambda^{-1}_{(\underset{\sim}{N})}\, x^k{}_{,K}\, N^K \tag{1.5.2}$$

where $\Lambda_{(\underset{\sim}{N})}$ is the stretch which is equal to the ratio ds/dS of the spatial to material arc length. Equation (1.5.2) determines how a material direction $\underset{\sim}{N}$ will rotate under the motion $\underset{\sim}{x}=\underset{\sim}{x}(\underset{\sim}{X},t)$. This is the same expression well-known in classical continuum mechanics (cf. Eringen, 1962, p. 21).

Consider now a unit vector $\underset{\sim}{H}=H^K\underset{\sim}{G}_K$ rigidly attached to the material particle at $\underset{\sim}{X}$ at $t=0$. This vector rotates to the vector $\underset{\sim}{\eta}$ at time t, where

$$\eta^k=\chi^k{}_K H^K \quad . \tag{1.5.3}$$

The quantity

$$\cos\psi = \underset{\sim}{n}\cdot\underset{\sim}{\eta} = \mathfrak{C}_{KL}N^K H^L/\Lambda_{(\underset{\sim}{N})} \quad . \tag{1.5.4}$$

is a measure of the difference between material elements rotating under the macromotion and the micromotion. If we take $\underset{\sim}{N}=\underset{\sim}{G}_K/\sqrt{G_{\underline{KK}}}$, $\underset{\sim}{H}=\underset{\sim}{G}_L/\sqrt{G_{\underline{LL}}}$ then (Fig. 1.5.2)

$$\underset{\sim}{n}_{(K)}\cdot\underset{\sim}{\eta}_{(L)} = \mathfrak{C}_{KL}/\Lambda_{(\underset{\sim}{N})}\sqrt{G_{\underline{KK}}G_{\underline{LL}}} \quad . \tag{1.5.5}$$

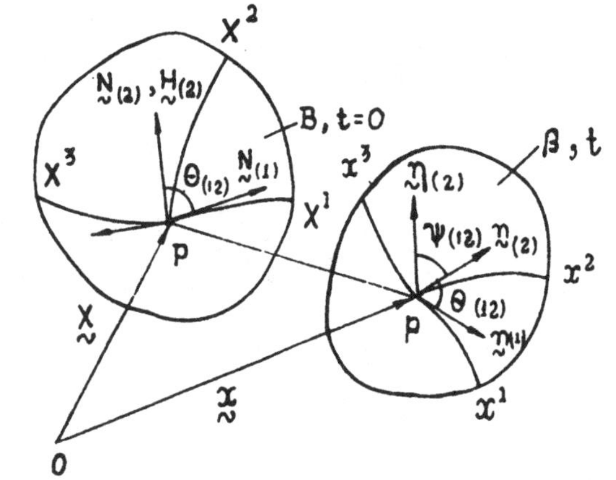

Fig. 1.5.2.
Change of vectors under the macro-and microbation.

This provides a geometrical significance for $\mathfrak{C}_{KL}$. The tangent vectors $\underset{\sim}{N}_{(1)}$, $\underset{\sim}{N}_{(2)}$ to the coordinate curves deform to the tangents $\underset{\sim}{n}_{(1)}$, $\underset{\sim}{n}_{(2)}$ of the deformed coordinate curves. However $\underset{\sim}{H}_{(2)}$, which is originally taken tangent to the X^2-coordinate curve, does not, in general, remain tangent to the x^2-curve. The angle $\psi_{(22)}$ between $\underset{\sim}{\eta}_{(2)}$ and $\underset{\sim}{\eta}_{(2)}$ is the result of the independent microrotation of the material particle. According to (1.5.5) the angle between $\underset{\sim}{\eta}_{(1)}$ and $\underset{\sim}{\eta}_{(2)}$ is proportional to $\mathcal{C}_{12}$. Similarly the angle between $\underset{\sim}{\eta}_{(1)}$ and $\underset{\sim}{\eta}_{(1)}$ is proportional to $\mathcal{C}_{11}$.

The wryness tensor Γ_{KL} is a measure of the relative orientation of two neighboring triads. This point may be seen by carrying (1.4.11) to (1.4.27) which results in (Kafadar and Eringen [1970])

$$\Gamma_{KL} \equiv N_K \Theta_{:L} + \sin\Theta N_{K:L} - (1-\cos\Theta)\varepsilon_{KMN} N^M N^N_{:L} \qquad (1.5.6)$$

where we wrote $N^K \equiv g^K{}_k n^k$, $\Theta \equiv \theta$. Note that for $\Theta \ll 1$ we obtain

$$\Gamma_{KL} \simeq \varphi_{K:L}\,, \quad \varphi_K \equiv g_K{}^k \varphi_k\,. \qquad (1.5.7)$$

This last linear approximation makes the assertion particularly clear.

When the direction $\underset{\sim}{N}$ is constant throughout B, $N_{K:L} = 0$ and (1.5.6) gives exactly

$$\Gamma_{KL} = \varphi_{K:L} \qquad (1.5.8)$$

so that $\det \Gamma_{KL} = 0$. This shows that Γ_{KL} contrary to $\mathcal{C}_{KL}$ may not posses an inverse.

The ratio of the deformed volume element dv to the undeformed element dV is given by $\dfrac{dv}{dV} = \sqrt{\det C^K{}_L} = J$

as in the classical theory, since the rotation of directors does not affect the length changes. But since

$$C_{KL} = d_{KM} d^{M}{}_{L} \tag{1.5.9}$$

we have

$$\frac{dv}{dV} = J = \det d^{K}{}_{L} = \sqrt{\det C^{K}{}_{L}} \ . \tag{1.5.10}$$

The change of an area element is calculated by

$$da_{k} = J X^{K}{}_{,k} \, dA_{K} \tag{1.5.11}$$

as in the classical theory, where da_k and dA_K are respectively the area vectors in the deformed and undeformed bodies.

1.6. Rate of Rotation and Strain Measures

The time rate of change of tensor fields associated with the material particles of the body is basic to the kinematics of continua.

Definition 1. The material time rate of a vector (or tensor) $\underset{\sim}{f}$ is defined as

$$\frac{d\underset{\sim}{f}}{dt} = \left.\frac{\partial \underset{\sim}{f}}{\partial t}\right|_{\underset{\sim}{X}} \tag{1.6.1}$$

where the subscript $\underset{\sim}{X}$ accompanying a vertical bar indicates that $\underset{\sim}{X}$ is held constant in the differentiation of $\underset{\sim}{f}$. If $\underset{\sim}{f}$ is a material function, for example,

$$\underset{\sim}{f} = \underset{\sim}{f}(\underset{\sim}{X}, t) = F^{K}(\underset{\sim}{X}, t)\,\underset{\sim}{G}_{K}$$

then it is clear that

$$\frac{d\underset{\sim}{f}}{dt} = \frac{\partial F^K}{\partial t}\underset{\sim}{G}_K \qquad (1.6.2)$$

since $\underset{\sim}{G}_K$ are functions of $\underset{\sim}{X}$ only. If on the other hand $\underset{\sim}{f}$ is a spatial function, for example,

$$\underset{\sim}{f} = \underset{\sim}{f}(\underset{\sim}{x}, t) = f^k(\underset{\sim}{x}, t)\, g_k(\underset{\sim}{x})$$

then

$$\frac{d\underset{\sim}{f}}{dt} = \left(\frac{\partial f^k}{\partial t}\Big|_{\underset{\sim}{x}} + \frac{\partial f^k}{\partial x^\ell}\frac{dx^\ell}{dt}\right)\underset{\sim}{g}_k + f^k\frac{\partial \underset{\sim}{g}_k}{\partial x^\ell}\frac{dx^\ell}{dt} . \qquad (1.6.3)$$

The partial derivative $\partial \underset{\sim}{g}_k / \partial x^\ell$ can be expressed as

$$\frac{\partial \underset{\sim}{g}_k}{\partial x^\ell} = \left\{ {m \atop k\ell} \right\} g_m \qquad (1.6.4)$$

where

$$\left\{ {m \atop k\ell} \right\} = g^{mr}[k\ell, r] = \frac{1}{2} g^{mr}(g_{kr,\ell} + g_{\ell r,k} - g_{k\ell,r})$$

is the <u>Christoffel symbol of the second kind</u>, $[\ell m, r]$ being the first kind. Substituting (1.6.4) into (1.6.3) we get

$$\frac{d\underset{\sim}{f}}{dt} = \frac{Df^k}{Dt}\underset{\sim}{g}_k \qquad (1.6.5)$$

where

$$\frac{Df^k}{Dt} = \frac{\partial f^k}{\partial t} + f^k{}_{;\ell}\,\dot{x}^\ell \qquad (1.6.6)$$

is the <u>material derivative</u> of f^k with

$$f^k{}_{;\ell} \equiv f^k{}_{,\ell} + \left\{ {k \atop \ell m} \right\} f^m \qquad (1.6.7)$$

being the covariant partial derivative of f^k.

Following (1.6.5) we have, for the velocity $\underset{\sim}{v}$ and acceleration $\underset{\sim}{a}$ of a material point in the deformed body,

(1.6.8) $$\underset{\sim}{v} = v^k \underset{\sim}{g}_k \,, \quad v^k \equiv \frac{\partial x^k}{\partial t}\Big|_{\underset{\sim}{X}} \equiv \dot{x}^k$$

(1.6.9) $$\underset{\sim}{a} = a^k \underset{\sim}{g}_k \,, \quad a^k \equiv \frac{Dv^k}{Dt} = \frac{\partial v^k}{\partial t} + v^k{}_{;\ell} v^\ell \,.$$

Two important lemmas for the calculation of the time rates of sundry tensors of micropolar continua are given below

Lemma 1. The material derivative of $x^k{}_{,K}$ is given by

(1.6.10) $$\frac{Dx^k{}_{,K}}{Dt} = v^k{}_{;\ell} x^\ell{}_{,K} \quad \text{or} \quad \frac{D}{Dt}(dx^k) = v^k{}_{;\ell} dx^\ell \,.$$

The proof of (1.6.10) follows from the fact D/Dt and $\partial/\partial X^K$ commute. Thus form

$$\frac{D}{Dt}\cdot(d\underset{\sim}{p}) = \frac{D}{Dt}(\underset{\sim}{x}_{,K}\, dX^K) = \underset{\sim}{v}_{,K}\, dX^K$$

but since

$$\underset{\sim}{v} = v^k(\underset{\sim}{x}, t)\underset{\sim}{g}_k(x)$$

$$\underset{\sim}{v}_{,K} = v^k{}_{,\ell}\, x^\ell{}_{,K} \underset{\sim}{g}_k + v^k \underset{\sim}{g}_{k,\ell}\, x^\ell{}_{,K}$$

or using (1.6.4)

(1.6.11) $$\underset{\sim}{v}_{,K} = v^k{}_{;\ell}\, x^\ell{}_{,K} \underset{\sim}{g}_k \,.$$

Consequently

$$\frac{D}{Dt}(d\underset{\sim}{p}) = v^{k}{}_{;\ell}\, dx^{\ell}\, \underset{\sim}{g}_{k} \tag{1.6.12}$$

where the coefficient of $\underset{\sim}{g}_k$ is $(1.6.10)_2$ and thus the proof of the theorem.

Corollary. The material time rate of $X^K{}_{,k}$ is given by

$$\frac{D}{Dt}(X^K{}_{,k}) = -v^{\ell}{}_{;k}\, X^K{}_{,\ell} \;. \tag{1.6.13}$$

This follows from differentiating

$$X^K{}_{,k}\, x^k{}_{,L} = \delta^K{}_L$$

and using this expression once again.

Definition. The microgyration tensor $\nu_{k\ell}$ is the angular velocity tensor of $\underset{\sim}{\chi}$ defined by, Eringen [1964] ,

$$\nu_{k\ell} \equiv \frac{D}{Dt}(\chi_{kK})\, \chi_{\ell}{}^{K} = -\nu_{\ell k} \;. \tag{1.6.14}$$

Since $\nu_{k\ell}$ is a skew-symmetric tensor in our three-dimensional space, we can introduce an equivalent axial vector ν_k, called the microgyration vector, by

$$\nu_k \equiv -\frac{1}{2}\,\varepsilon_{k\ell m}\, \nu^{\ell m} \quad , \quad \nu_{k\ell} = -\varepsilon_{k\ell m}\, \nu^{m} \;. \tag{1.6.15}$$

From the orthogonality of $\underset{\sim}{\chi}$ it follows that

$$\frac{D}{Dt}(\chi^k{}_K) = \nu^k{}_{\ell}\, \chi^{\ell}{}_K \;. \tag{1.6.16}$$

Thus

Lemma 2. The material time rate of $\underset{\sim}{\chi}$ is given by (1.6.16). Similarly we have

(1.6.17) $$\frac{D}{Dt}\overset{-1}{\chi}{}^{K}{}_{k} = -\nu^{\ell}{}_{k}\overset{-1}{\chi}{}^{K}{}_{\ell} \; .$$

This follows from differentiating (1.4.5) and using (1.6.16).

Theorem 1. The microgyration vector is related to the time rate of the rotation vector, φ_k by (Kafadar and Eringen [1970])

(1.6.18) $$\nu_k = \Lambda_{k\ell}\dot{\varphi}^{\ell}$$

where

(1.6.19) $$\Lambda_{k\ell} \equiv \frac{\sin\theta}{\theta} g_{k\ell} - \frac{1-\cos\theta}{\theta^2}\varepsilon_{k\ell m}\varphi^m + \left(1 - \frac{\sin\theta}{\theta}\right)\frac{1}{\theta^2}\varphi_k\varphi_\ell$$
$$\theta \equiv \sqrt{\varphi_k\varphi^k} \; .$$

The proof of this follows from taking the time rate of (1.4.11) and using (1.6.15) and (1.6.16).

The inverse of (1.6.18) exists and is given by

(1.6.20) $$\dot{\varphi}_\ell = \overset{-1}{\Lambda}_{\ell k}\nu^k$$

where

(1.6.21) $$\overset{-1}{\Lambda}_{\ell k} = \frac{\theta}{2}\cot\frac{\theta}{2} g_{k\ell} + \frac{1}{2}\varepsilon_{\ell k m}\varphi^m + \left(1 - \frac{\theta}{2}\cot\frac{\theta}{2}\right)\frac{1}{\theta^2}\varphi_k\varphi_\ell \; .$$

When $\theta << 1$ the linear approximation to $\underset{\sim}{\Lambda}$ and $\overset{-1}{\underset{\sim}{\Lambda}}$ are

(1.6.22a) $$\Lambda_{k\ell} \simeq g_{k\ell} - \frac{1}{2}\varepsilon_{k\ell m}\varphi^m$$

$$\overset{-1}{\Lambda}_{\ell k} \simeq g_{\ell k} + \frac{1}{2}\varepsilon_{\ell k m}\varphi^m \tag{1.6.22b}$$

so that (1.6.18) reduces to

$$\nu_k \simeq \dot{\varphi}_k \tag{1.6.23}$$

a result already used by Eringen in the linear theory in 1964.

<u>Theorem. The material derivative of the Cosserat deformation tensor is given by</u>

$$\frac{D\mathcal{C}_{KL}}{Dt} = (v_{\ell;k} + \nu_{k\ell})\, x^k_{,K}\, \chi^\ell_{\ \,L} \;. \tag{1.6.24}$$

To prove this, we differentiate (1.4.26) and use (1.6.10) and (1.6.16). i.e.,

$$\frac{D\mathcal{C}_{KL}}{Dt} = \frac{Dx^k_{,K}}{Dt}\chi_{kL} + x^k_{,K}\frac{D}{Dt}\chi_{kL} = v^k_{\ ;\ell}\, x^\ell_{,K}\,\chi_{kL} + x^k_{,K}\,\nu_{k\ell}\,\chi^\ell_{\ \,L} \;.$$

This is the same as (1.6.24).

Similarly with a little more manipulation, we can show that

$$\frac{D\Gamma_{KL}}{Dt} = \nu_{k;\ell}\, x^\ell_{,L}\, \chi^k_{\ \,K} \tag{1.6.25}$$

$$\frac{D c_{k\ell}}{Dt} = v_{k;m}\, c^m_{\ \,\ell} + c_k^{\ m}\, \nu_{\ell m} \tag{1.6.26}$$

$$\frac{D\gamma_{k\ell}}{Dt} = \nu_{km}\gamma^m_{\ \,\ell} + \gamma_{km}\nu_\ell^{\ n} + \nu_{k;m}\, c^m_{\ \,\ell} \;. \tag{1.6.27}$$

A motion is called rigid if $D\mathfrak{C}_{KL}/Dt = D\Gamma_{KL}/Dt = 0$.

Theorem : A motion is rigid if and only if

(1.6.28) $$v_{\ell;k} + v_{k\ell} = 0 \; , \quad v_{k;\ell} = 0 \; .$$

The proof is obvious from (1.6.24) and (1.6.25).

The symmetric part of (1.6.28) gives

$$v_{(k;\ell)} = 0$$

whose general solution is

$$z^k = R^k{}_K(t) Z^K + b^k(t)$$

where $R^k{}_K$ is orthogonal and is the classical rotation tensor. But $D\mathfrak{C}_{KL}/Dt = 0$ gives $\mathfrak{C}_{KL} = \mathfrak{C}_{KL}\big|_{t=0} = G_{KL}$ so that

$$z^k{}_{,K}\chi_{KL} = G_{KL} \text{ or } \chi_k{}^K = Z^K{}_{,k} = R_k{}^K(t) \; .$$

Thus the general solution of (1.6.28) is

(1.6.29) $$z^k(\underset{\sim}{Z}, t) = R^k{}_K(t) Z^K + b^k(t)$$

$$\chi^k{}_K(\underset{\sim}{Z}, t) = R^k{}_K(t) \; .$$

For a rigid motion we have

(1.6.30) $$\mathfrak{C}_{KL} = G_{KL} \; , \; \Gamma_{KL} = 0 \; .$$

1.7. Compatibility Conditions

For a given set of micropolar strain tensors $\mathscr{C}_{KL}, \Gamma_{KL}$ we have 18 first order partial differential equations to determine the six unkowns x^k and φ^k. For the single-valued displacement and rotation fields to exist, certain compatibility conditions must be satisfied (Kafadar and Eringen, 1970).

Theorem. For a simply connected body, the necessary and sufficient conditions for $\mathscr{C}_{KL}$ and Γ_{KL} to be integrable are :

$$\varepsilon^{KMN}\left(\mathscr{C}^{L}_{M:N} + \varepsilon^{LPQ}\Gamma_{PN}\mathscr{C}_{MQ}\right) = 0 \tag{1.7.1}$$

$$\varepsilon^{KMN}\left(\Gamma^{L}_{N:M} + \frac{1}{2}\varepsilon^{LPQ}\Gamma_{PM}\Gamma_{QN}\right) = 0 . \tag{1.7.2}$$

For a simply connected body B, the necessary and sufficient conditions that $x^k{}_{,K}$ and $\chi^k{}_{K:L}$ be integrable are

$$x^k{}_{:KL} = x^k{}_{:LK} , \quad \chi^k{}_{K:LM} = \chi^k{}_{K:ML} .$$

But from the expression (1.4.26) of $\mathscr{C}_{KL}$ we have

$$x^k{}_{,K} = \mathscr{C}_{KN}\chi^{kN} .$$

Thus

$$x^k{}_{:KL} = \mathscr{C}_{KN:L}\chi^{kN} + \mathscr{C}_{KN}\chi^{kN}{}_{:L} = \mathscr{C}_{LN:K}\chi^{kN} + \mathscr{C}_{LN}\chi^{kN}{}_{:K}$$

or

$$\mathscr{C}_{KM:L} - \mathscr{C}_{LM:K} = \mathscr{C}_L{}^N\varepsilon_{NMP}\Gamma^P{}_K - \mathscr{C}_K{}^N\varepsilon_{NMP}\Gamma^P{}_L$$

or

$$\varepsilon^{RKL}\left(\sigma_{K\,:L}^{\;M} + \varepsilon^{MPQ}\,\Gamma_{PL}\,\sigma_{KQ}\right) = 0$$

which is identical to (1.7.1). Equation $(1.7.2)_2$ follows analogously by considering $\Gamma_{KL:M}$.

From (1.7.1) we can solve for Γ_{KL}:

(1.7.3) $$\Gamma_{KL} = J^{-1}\varepsilon^{PMN}\,\sigma_{M:N}^{\;Q}\left(\frac{1}{2}\,\sigma_{LK}\sigma_{PQ} - \sigma_{PK}\sigma_{LQ}\right)$$

which provides a connection between $\underset{\sim}{\Gamma}$ and the tensor $\underset{\sim}{\sigma}$ and its gradient.

For the linear theory (1.7.1) reduce to

(1.7.4) $$\sigma_{KM:L} - \sigma_{LM:K} - \varepsilon_{LMP}\Gamma^{P}_{\ K} + \varepsilon_{KMP}\Gamma^{P}_{\ L} = 0$$

(1.7.5) $$\varepsilon^{KMN}\,\Gamma^{L}_{\ N:M} = 0$$

of which (1.7.4), in a different notation, was given by Sandru [1966] , and both were also obtained by Eringen [1967 c] .

Chapter 2
BALANCE LAWS

2.1. Scope of the Chapter

This chapter is devoted to the development of the global and local balance laws. The general forms of the balance laws and jump conditions at a moving discontinuously surface sweeping through the body are given in Art. 2.2. The conservation of mass and microinertia are presented in Art. 2.3., the balance of momenta in Art. 2.4. The final section (Art. 2.5) of this chapter gives the expression of the local balance of energy and entropy and their associated jump conditions.

2.2. General Form of Balance Laws

The balance laws are the expression of the balance of the time rate of change of a quantity with sources in the body and an influx through its surface. Thus if ψ is a tensor field per unit volume, g its body source, and $\underset{\sim}{\tau}$ is the flux per unit area contributing to the change of ψ, then the general form of any balance law, over the material body having volume V enclosed within a material surface S, at time t, is

$$\frac{D}{Dt}\int_{V-\sigma}\psi\, dv - \oint_{S-\sigma}\tau^k da_k - \int_{V-\sigma} g\, dv = 0 . \qquad (2.2.1)$$

To include the possibility that there may exist a surface of dis-

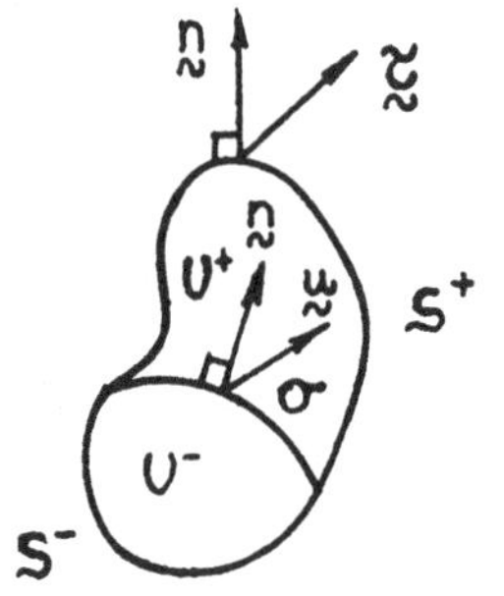

Fig. 2.2.1.
Body swept by a discontinuity surface σ,
$V-\sigma \equiv V^{+}+V^{-}$, $S-\sigma \equiv S^{+}+S^{-}$

continuity σ moving across the body with a velocity $\mathbf{u}$ (generally different from the velocity $\mathbf{v}$ of the material point) in the direction of its positive unit normal $\mathbf{n}$,(Fig.2.2.1), we write (2.2.1) only over those parts of the body which exclude σ.

It can,be shown that Eringen [1967, App. A]

$$(2.2.2)\qquad \frac{D}{Dt}\int_{V-\sigma}\psi\, dv = \int_{V-\sigma}\left[\frac{\partial\psi}{\partial t}+\nabla\cdot(\psi\mathbf{v})\right]dv + \int_{\sigma}[\![\psi(\mathbf{v}-\mathbf{u})]\!]\cdot d\mathbf{a}$$

$$(2.2.3)\qquad \oint_{S-\sigma}\boldsymbol{\tau}\cdot d\mathbf{a} = \int_{V-\sigma}\nabla\cdot\boldsymbol{\tau}\, dv + \int_{\sigma}[\![\boldsymbol{\tau}]\!]\cdot d\mathbf{a}$$

of which the last one is the generalized Green-Gauss theorem, Eringen 1962, App. . Here a boldface bracket $[\![\;]\!]$ indicates the jump of the enclosure as σ is approached from the front side of its positive normal and the back.

Employing the integral relations, (2.2.1) can be transformed into

$$(2.2.4)\qquad \int_{V-\sigma}\left[\frac{\partial\psi}{\partial t}+\nabla\cdot(\psi\mathbf{v})-\nabla\cdot\boldsymbol{\tau}-g\right]dv + \int_{\sigma}[\![\psi(\mathbf{v}-\mathbf{u})-\boldsymbol{\tau}]\!]\cdot d\mathbf{a} = 0.$$

If it is postulated that (2.2.4) is valid for every part of the body, then we obtain the <u>local forms</u> of the general balance law

$$(2.2.5)\qquad \frac{\partial\psi}{\partial t}+\nabla\cdot(\psi\mathbf{v})-\nabla\cdot\boldsymbol{\tau}-g=0 \qquad \text{in } V-\sigma$$

$$[\underset{\sim}{\psi} \cdot (\underset{\sim}{v} - \underset{\sim}{u}) - \underset{\sim}{\tau}] \cdot \underset{\sim}{n} = 0 \text{ on } \sigma \ . \tag{2.2.6}$$

The first of these equations localizes the balance law and the second expresses the local jump conditions on the discontinuity surface. As we shall see, all local balance laws of continuum physics have the forms (2.2.5) and (2.2.6). In fact, even the second law of thermodynamics has this form except that the equality sign "=", in this case must be replaced by the sign " $\geqq$ ".

2.3. Conservation of Mass and Microinertia

Elements of the body, called material particles, are posited to have two basic physical properties, namely, mass and microinertia. This is motivated from our experience with rigid bodies. The material particles of a micropolar body phenomenologically are considered to be small rigid bodies. In the spirit of continuum mechanics, we assume that both mass and microinertia are continuous fields so that a positive mass density, ρ_0 , and a positive definite symmetric microinertia tensor density, I_{KL} , exist at each point of $\mathcal{B}$. Since the inertia tensor is a measure of the orientation of the material particles, it must therefore change with the micromotion. According to Eringen and Suhubi [1964] the transformation law governing I_{KL} is

$$i^{kl} = \chi^k{}_K \chi^l{}_L I^{KL} \tag{2.3.1}$$

where $i^{\cdot k\ell}$ is the spatial microinertia tensor. Here we may consider this to be the definition of $i^{\cdot k\ell}$ if we wish. It is derivable from the consideration of the deformation of micro-elements.

Related to the microinertia tensors I^{KL} and $i^{\cdot k\ell}$ are the material and spatial microinertia tensors J^{KL} and $j^{\cdot k\ell}$ which may prove to be more convenient on some occasions. These are defined by

(2.3.2) $$J^{KL} \equiv I^{M}{}_{M} G^{KL} - I^{KL}$$

(2.3.3) $$j^{\cdot k\ell} \equiv i^{\cdot m}{}_{m}\, g^{k\ell} - i^{\cdot k\ell} .$$

Since $\underset{\sim}{\chi}$ is orthogonal, from these and (2.3.1) it follows that

(2.3.4) $$j^{\cdot k\ell} \equiv \chi^{k}{}_{K}\, \chi^{\ell}{}_{L}\, J^{KL} .$$

Fundamental Principle I (Law of Conservation of Mass) : The mass of the body is unchanged with motion. Mathematically

(2.3.5) $$\int_{V_0-\sigma_0} \rho_0 \, dV = \int_{V-\sigma} \rho \, dv$$

V_0 is the original volume of the body at $t=0$ and V is that at any other time t and ρ_0 and ρ are the respective mass densities. Since we have $dv = J\,dV$, we also have

(2.3.6) $$\int_{V_0-\sigma} (\rho_0 - \rho J)\, dV = 0 \quad , \quad J \equiv \sqrt{g/G}\, \det(x^{k}{}_{,K})$$

where $g \equiv \det g_{k\ell}$ and $G \equiv \det G_{KL}$. Alternatively, by differentiation of (2.3.5), we have equivalently

$$\frac{D}{Dt}\int_{V-\sigma} \rho \, dv = 0 \, . \tag{2.3.7}$$

Comparing this form with the general balance law we see that $\psi=\rho$ $\tau=0$, $g=0$. Thus if the mass is conserved locally then (2.3.7), according to (2.2.5) and (2.2.6), gives

$$\frac{\partial\rho}{\partial t} + \nabla\cdot(\rho \mathbf{v}) = 0 \quad , \quad V-\sigma \tag{2.3.8}$$

$$[\rho(\mathbf{v}-\mathbf{u})]\cdot\mathbf{n} = 0 \quad , \quad \text{on } \sigma \ . \tag{2.3.9}$$

These are the local forms of the law of conservation of mass. An alternative equivalent local form, replacing (2.3.8) is

$$\frac{D}{Dt}(\rho \, dv) = 0 \quad , \quad \text{in } V-\sigma \quad . \tag{2.3.10}$$

If we wish to employ (2.3.6), the corresponding local form is

$$\rho_0 - \rho J = 0 \quad , \text{in } V-\sigma \tag{2.3.11}$$

whose material derivative gives (2.3.8).

Fundamental Principle II (Law of Conservation of Microinertia) : The microinertia of the body is unchanged with the motion.

$$\int_{V_0-\sigma_0} \rho_0 \mathbf{I} \, dV = \int_{V-\sigma} \rho \, i^{kl} \boldsymbol{\chi}_k \boldsymbol{\chi}_l \, dv \tag{2.3.12}$$

where $\boldsymbol{\chi}_k = \chi_k{}^K \mathbf{G}_K$. Alternatively, by taking the material time rate, we write

$$\frac{d}{dt}\int_{V-\sigma} \rho \, i^{kl} \boldsymbol{\chi}_k \boldsymbol{\chi}_l \, dv \quad . \tag{2.3.13}$$

Note that (2.3.12) and (2.3.13) are not written in component form because we are not permitted to add components of tensors associated with different parts of the body.

If it is postulated that this law is valid for every part of the body, following the scheme of (2.2.5) and (2.2.6), we obtain

$$\frac{D}{Dt}\left(\dot{i}^{k\ell}\chi_k{}^K\chi_\ell{}^L\right)=0 \quad \text{in } V-\sigma$$

$$\left[\dot{i}^{k\ell}\chi_k{}^K\chi_\ell{}^L(\underset{\sim}{v}-\underset{\sim}{u})\right]\cdot\underset{\sim}{n}=0 \quad \text{on } \sigma$$

Expanding the first of these and using (1.6.17) we have

(2.3.14) $$\frac{D\dot{i}^{k\ell}}{Dt}-\dot{i}^{kr}v^{\ell}{}_{,r}-\dot{i}^{r\ell}v^{k}{}_{,r}=0 \quad \text{in } V-\sigma$$

(2.3.15) $$\left[\underset{\sim}{i}^{k\ell}\chi_k{}^K\chi_\ell{}^L(\underset{\sim}{v}-\underset{\sim}{u})\right]\cdot\underset{\sim}{n}=0 \quad \text{on } \sigma .$$

These are the expressions of the <u>local conservation of microinertia</u> and <u>the jump conditions</u>, given first by Eringen [1964 b].

If we wish, we may employ $\dot{j}^{k\ell}$ instead of $\dot{i}^{k\ell}$. The forms of these equations remain unchanged.

(2.3.16) $$\frac{D\dot{j}^{k\ell}}{Dt}-\dot{j}^{kr}v^{\ell}{}_{,r}-\dot{j}^{r\ell}v^{k}{}_{,r}=0 \text{ in } V-\sigma$$

(2.3.17) $$\left[\dot{\underset{\sim}{j}}^{k\ell}\chi_k{}^K\chi_\ell{}^L(\underset{\sim}{v}-\underset{\sim}{u})\right]\cdot\underset{\sim}{n}=0 \quad \text{on } \sigma .$$

Just as in the continuity equations (2.3.8) and (2.3.9), the local conservation laws (2.3.14) and (2.3.15) of microinertia are an integral part of the theory of micropolar continua.

2.4. Balance of Momenta

With each particle of a micropolar body, there is associated a momentum density $\rho \underset{\sim}{v}$, where $\underset{\sim}{v}$ is the velocity vector and a spin density, $\rho \underset{\sim}{\sigma}$. The spin density arises from the intrinsic rotatory momentum of the element. It is defined as

$$\sigma^k \equiv j^{k\ell} \nu_\ell \quad . \tag{2.4.1}$$

The total moment of momentum of an element of the micropolar body is therefore given by $\underset{\sim}{p} \times \rho \underset{\sim}{v} + \rho \underset{\sim}{\sigma}$.

Fundamental Principle III (Law of Balance of Momentum) : The time rate of change of the total momentum of a micropolar body is equal to the sum of surface and body forces acting on the body. Mathematically,

$$\frac{d}{dt} \int_{V-\sigma} \rho \underset{\sim}{v} \, dv = \oint_{S-\sigma} \underset{\sim}{t}_{(\underset{\sim}{n})} da + \int_{V-\sigma} \rho \underset{\sim}{f} \, dv \quad . \tag{2.4.2}$$

Here $\underset{\sim}{t}_{(\underset{\sim}{n})}$ is the surface traction acting on the surface of the body whose exterior unit normal is $\underset{\sim}{n}$, and $\underset{\sim}{f}$ is the body force density.

If (2.4.1) is postulated to be valid for all parts of the body, we obtain the localization of the balance of momentum. In this case, by applying (2.4.1) to a tetrahedral volume element, Δv enclosed within three coordinate surfaces and the forth one being the element of area da lying on the surface of the body, in the limit $\Delta v \to 0$, we obtain (cf. Eringen, 1962 ch.III)

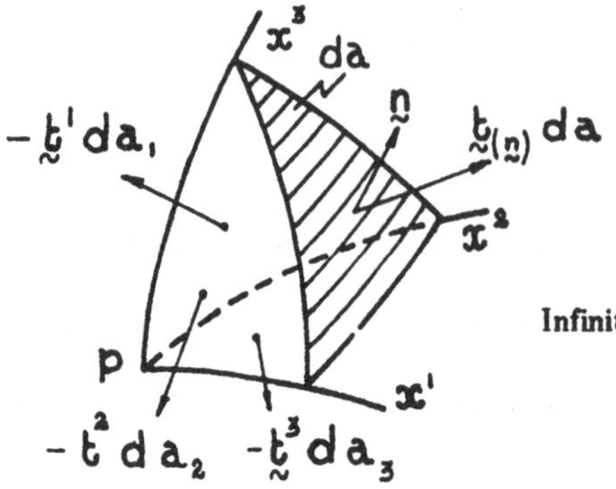

Fig. 2.4.1

Infinitesimal Tetrahedron near the surface of the body.

$$t_{(n)} \, da = t^k \, da_k \tag{2.4.3}$$

where $-t^k$ are the surface tractions on the coordinate surfaces x^k = const. Since the tetrahedron surface is closed we have

$$da_k = n_k \, da \tag{2.4.4}$$

where n is the unit external normal to the surface of the body.

Combining (2.4.4) and (2.4.5) we get

$$t_{(n)} = t^k n_k \, . \tag{2.4.5}$$

Since t^k are independent of n it follows that

$$t_{(-n)} = -t_{(n)} \, . \tag{2.4.6}$$

This is the mathematical expression of the fact that the action of one part of the body through a surface on the other part is equal and opposite to the action of the latter upon the former. This is the extension of the third law of Newton to continuum mechanics.

Definition. The stress tensor $t^{k\ell}$ is the ℓth component of the traction t^k, i.e.,

$$t^k = t^{k\ell} g_\ell \quad , \quad t^{k\ell} \equiv t^k \cdot g^\ell \, . \tag{2.4.7}$$

Substituting (2.4.2) into (2.4.1) gives

$$\frac{d}{dt}\int_{V-\sigma}\rho\,\underset{\sim}{v}\,dv-\oint_{S-\sigma}\underset{\sim}{t}^{k}da_{k}-\int_{V-\sigma}\rho\underset{\sim}{f}\,dv=\underset{\sim}{0}\;. \qquad (2.4.8)$$

Applying the general scheme (2.2.5) to (2.4.8) and the local conservation of mass, we obtain the local balance of momentum.

$$\frac{1}{\sqrt{g}}\frac{\partial}{\partial x^{k}}(\sqrt{g}\,\underset{\sim}{t}^{k})+\rho(\underset{\sim}{f}-\dot{\underset{\sim}{v}})=\underset{\sim}{0} \quad \text{in} \quad V-\sigma \qquad (2.4.9)$$

$$[\underset{\sim}{t}^{k}-\rho\underset{\sim}{v}(v^{k}-u^{k})]n_{k}=\underset{\sim}{0} \quad \text{on } \sigma \qquad (2.4.10)$$

In (2.4.8) the first term is the expression of divergence for the vector field $\underset{\sim}{t}^{k}$, written in compact form.

In terms of the components of the tensors involved, (2.4.9) and (2.4.10) read

$$t^{kl}{}_{;k}+\rho(f^{l}-\dot{v}^{l})=0 \quad \text{in} \quad V-\sigma \qquad (2.4.11)$$

$$[t^{kl}-\rho v^{l}(v^{k}-u^{k})]n_{k}=0 \quad \text{on } \sigma \qquad (2.4.12)$$

which are obtained by using (2.4.7) and the differentiation formulas such as (1.6.4).

Equations (2.4.11) and (2.4.12) are the expression of the <u>first law of Cauchy</u>.

In a micropolar body, in addition to body forces the existence of an independent body couple density is postulated. The volume elements of the body can thus sustain a body force $\underset{\sim}{f}$ and body couple $\underset{\sim}{l}$ per unit mass. Through any surface of an element of the body, the effect of the contacting surface of the remaining part of the body is transmitted by the surface traction $\underset{\sim}{t}_{(\underset{\sim}{n})}$ and surface couple $\underset{\sim}{m}_{(\underset{\sim}{n})}$. Thus the resultant couple per unit area of a surface element is given by

$\underset{\sim}{m}_{(\underset{\sim}{n})} + \underset{\sim}{p} \times \underset{\sim}{t}_{(\underset{\sim}{n})}$.

Fundamental Principle IV (Law of Balance of Moment of Momentum) : The time rate of change of the total moment of momentum is equal to the sum of all surface and body couples acting on the body. Mathematically,

$$(2.4.13)\quad \frac{d}{dt}\int_{V-\sigma}(\rho\underset{\sim}{\sigma}+\underset{\sim}{p}\times\rho\underset{\sim}{v})dv=\int_{S-\sigma}(\underset{\sim}{m}_{(\underset{\sim}{n})}+\underset{\sim}{p}\times\underset{\sim}{t}_{(\underset{\sim}{n})})da+\int_{V-\sigma}\rho(\underset{\sim}{\ell}+\underset{\sim}{p}\times\underset{\sim}{f})dv.$$

Similar to what is done for the tractions, by applying this law to an infinitesimal tetrahedron near the surface of the body, Fig. 2.4.1, in the limit as the volume of the tetrahedron approaches zero, we obtain

$$(2.4.14)\qquad \underset{\sim}{m}_{(\underset{\sim}{n})}da = \underset{\sim}{m}^k da_k$$

where $-\underset{\sim}{m}^k$ are the three surface couple vectors acting on the coordinate surfaces of the tetrahedron. Using (2.4.3) this gives

$$(2.4.15)\qquad \underset{\sim}{m}_{(\underset{\sim}{n})} = \underset{\sim}{m}^k n_k .$$

Since $\underset{\sim}{m}^k$ are independent of the unit normal $\underset{\sim}{n}$ of the surface, we have

$$(2.4.16)\qquad \underset{\sim}{m}_{(-\underset{\sim}{n})} = -\underset{\sim}{m}_{(\underset{\sim}{n})} .$$

This expresses the fact that the surface couple due to one part of the body acting on the other through the surface is equal and opposite to the latter acting upon the former.

Definition. The couple stress tensor $m^{k\ell}$ is the ℓth component of the surface couple $\underset{\sim}{m}^k$ i.e.,

$$\underset{\sim}{m}^k = m^{k\ell}\,\underset{\sim}{g}_\ell \ , \quad m^{k\ell} \equiv \underset{\sim}{m}^k \cdot \underset{\sim}{g}^\ell \ . \qquad (2.4.17)$$

Substituting (2.4.16) into (2.4.12) we have

$$\frac{d}{dt}\int_{V-\sigma} (\rho\underset{\sim}{\sigma} + \underset{\sim}{p}\times\rho\underset{\sim}{v})\,dv = \oint_{S-\sigma} (\underset{\sim}{m}^k + \underset{\sim}{p}\times\underset{\sim}{t}^k)\,da_k + \int_{V-\sigma}\rho(\underset{\sim}{\ell} + \underset{\sim}{p}\times\underset{\sim}{f})\,dv \qquad (2.4.18)$$

We now apply the general scheme (2.2.5), under the postulate that (2.4.18) is valid for every part of the body and obtain

$$\frac{1}{\sqrt{g}}\frac{\partial}{\partial x^k}(\sqrt{g}\,\underset{\sim}{m}^k) + \underset{\sim}{g}_k\times\underset{\sim}{t}^k + \rho(\underset{\sim}{\ell} - \dot{\underset{\sim}{\sigma}}) = \underset{\sim}{0} \quad \text{in } V-\sigma \qquad (2.4.19)$$

$$[\underset{\sim}{m}^k - \rho\underset{\sim}{\sigma}(v^k - u^k)]\,n_k = \underset{\sim}{0} \quad \text{on } \sigma \qquad (2.4.20)$$

where we used (2.3.8), (2.3.9), (2.4.9) and (2.4.10).

Equation (2.4.19) is the expression of the local balance of moment of momentum and (2.4.20) is the associated jump condition.

Upon substituting (2.4.16) and carrying out the differentiations we obtain the component forms

$$m^{k\ell}{}_{;k} + \varepsilon^{\ell mn}\, t_{mn} + \rho(\ell^\ell - \dot{\sigma}^\ell) = 0 \quad \text{in } V-\sigma \qquad (2.4.21)$$

$$[m^{k\ell} - \rho\sigma^\ell(v^k - u^k)]\,n_k = 0 \quad \text{on } \sigma \ . \qquad (2.4.22)$$

These are the expressions of the generalization of Cauchy's second law to a micropolar body.

2.5. Principles of Conservation of Energy and Entropy Inequality

Fundamental Principle V (Conservation of Energy) : The time rate of change of the sum of internal energy E and kinetic energy K is equal to the sum of the energy of surface and body loads $\mathcal{W}$ the heat energy and the body energy source, $\mathcal{Q}$ per unit time. Mathematically

$$\dot{E} + \dot{K} = \mathcal{W} + \mathcal{Q} . \tag{2.5.1}$$

For the requisite energy terms we have

$$E \equiv \int_{V-\sigma} \rho \varepsilon \, dv$$

$$\begin{aligned} K &\equiv \int_{V-\sigma} \left(\tfrac{1}{2} \rho v_k v^k + \tfrac{1}{2} \rho j_{k\ell} \nu^k \nu^\ell\right) dv \\ \mathcal{W} &\equiv \oint_{S-\sigma} (t^{k\ell} v_\ell + m^{k\ell} \nu_\ell) \, da_k + \int_{V-\sigma} \rho (f^\ell v_\ell + \ell^\ell \nu_\ell) \, dv \\ \mathcal{Q} &\equiv \oint_{S-\sigma} q^k \, da_k + \int_{V-\sigma} \rho h \, dv \end{aligned} \tag{2.5.2}$$

where ε is the internal energy density per unit mass, q^k is the heat vector on S , and h is the energy source (e.g., heat source).

If (2.5.1) is postulated to be valid for all parts of the body, then we say that the energy is conserved locally. In this case, applying the general scheme (2.2.5) and using the equations of balance of mass (2.3.8), (2.3.9), the microinertia (2.3.16), (2.3.17) and momenta, (2.4.11), (2.4.12), (2.4.21),

(2.4.22), we obtain

$$\rho\dot{\varepsilon} = t^{k\ell}(v_{\ell;k} + \nu_{k\ell}) + m^{k\ell}\nu_{\ell;k} + q^{k}{}_{;k} + \rho h \quad , \quad \text{in} \quad V-\sigma \tag{2.5.3}$$

$$[t^{k\ell}v_\ell + m^{k\ell}\nu_\ell + q^k - (\rho\varepsilon + \tfrac{1}{2}\rho \underset{\sim}{v}\cdot\underset{\sim}{v} + \tfrac{1}{2}\rho\underset{\sim}{\sigma}\cdot\underset{\sim}{\nu})(v^k - u^k)]n_k = 0, \text{ on } \sigma. \tag{2.5.4}$$

These are the expressions of the local conservation of energy and the local jump conditions for the energy on a moving discontinuity surface.

With each element of the body we associate a scalar called the entropy which is independent of mechanical variables. We assume that it is continuous and possesses continuous partial derivatives with respect to the variable x^k and time t (except on σ). Thus an entropy density, η, exists. In a micropolar body, a volume source entropy b and a surface flux of entropy vector $\underset{\sim}{S}$ are postulated to exist also. The total entropy flux Γ of the body is given by

$$\Gamma \equiv \int_{V-\sigma} \rho\gamma dv = \frac{D}{Dt}\int_{V-\sigma} \rho\eta\, dv - \int_{V-\sigma} \rho b\, dv - \oint_{S-\sigma} s^k da_k \, . \tag{2.5.5}$$

Fundamental Axiom VI (The second Law of Thermodynamics) : the total entropy flux of a micropolar body is non-negative for all thermomechanical processes. Mathematically

$$\Gamma \geq 0 \, . \tag{2.5.6}$$

When (2.5.6) is postulated to be valid for all parts of the body, we say that the entropy inequality is valid locally. In this case (2.5.5) gives

(2.5.7) $$\rho\gamma \equiv \rho\dot{\eta} - \rho b - s^k_{;k} \geq \quad \text{in } \mathcal{V} - \sigma$$

(2.5.8) $$[\rho\,\eta(\underset{\sim}{v} - \underset{\sim}{u}) - \underset{\sim}{S}]\cdot\underset{\sim}{n} \geq 0 \quad \text{on } \sigma \;.$$

So far, the local entropy law (2.5.7) and (2.5.8) has no relation to the thermomechanical quantities introduced in balance laws. The connection is established by identifying the entropy source b and the surface entropy flux $\underset{\sim}{S}$. To this end, we introduce the absolute temperature θ, which is postulated to be

(2.5.9) $$\theta > 0 \;, \quad \inf \theta = 0 \;.$$

Dimensions of $\underset{\sim}{S}$ and b (and consequently η) are fixed by writing

(2.5.10) $$\underset{\sim}{S} = \frac{\underset{\sim}{q}}{\theta} + \underset{\sim}{S}_1 \quad , \quad b = \frac{h}{\theta} + b_1$$

which present no restriction on their forms. For certain thermomechanical changes it may turn out that $\underset{\sim}{S}_1 = \underset{\sim}{0}$ and $b_1 = 0$. This seems to be the case in all classical theories of a single continuum. If in a body $\underset{\sim}{S}_1 = \underset{\sim}{0}$, $b_1 = 0$, we say that the body is simple thermomechanically.

We postulate that micropolar bodies are simple thermomechanically*). Upon substituting $\underset{\sim}{q}/\theta$ for $\underset{\sim}{S}$, and h/θ

---

*) For examples of bodies that are not simple thermomechanically, we mention mixtures and chemically reacting media (cf. Ingram and Eringen [1965])

for b, (2.5.7) and (2.5.8) become

$$\rho\gamma \equiv \rho\dot{\eta} - \rho\theta^{-1}h - (q^k/\theta)_{;k} \geq 0 \quad \text{in } V-\sigma \tag{2.5.11}$$

$$[\rho\eta(\underset{\sim}{v}-\underset{\sim}{u}) - \underset{\sim}{q}\,\theta^{-1}]\cdot\underset{\sim}{n} \geq 0 \quad \text{on } \sigma . \tag{2.5.12}$$

If we now eliminate h between (2.5.11) and (2.5.3), we obtain

$$\rho\gamma \equiv \rho(\dot{\eta} - \frac{\dot{\varepsilon}}{\theta}) + \frac{1}{\theta}t^{k\ell}(v_{\ell;k} + v_{k\ell}) + \frac{1}{\theta}m^{k\ell}v_{\ell;k} + \frac{q^k}{\theta^2}\theta_{,k} \geq 0 \text{ in } V-\sigma . \tag{2.5.13}$$

An expression that is more convenient for some purposes is obtained by introducing <u>Helmholtz' free energy</u>

$$\psi \equiv \varepsilon - \theta\eta . \tag{2.5.14}$$

With this we obtain

$$\rho\gamma \equiv -\frac{\rho}{\theta}(\dot{\psi} + \dot{\theta}\eta) + \frac{1}{\theta}t^{k\ell}(v_{\ell;k} + v_{k\ell}) + \frac{1}{\theta}m^{k\ell}v_{\ell;k} + \frac{q^k}{\theta^2}\theta_{,k} \geq 0 \text{ in } V-\sigma \tag{2.5.15}$$

$$[\rho\eta(\underset{\sim}{v}-\underset{\sim}{u}) - \frac{\underset{\sim}{q}}{\theta}]\cdot\underset{\sim}{n} \geq 0 \quad \text{on } \sigma . \tag{2.5.16}$$

For the total entropy flux we have

$$\Gamma \equiv \frac{D}{Dt}\int_{V-\sigma}\rho\eta\, dv - \oint_{S-\sigma}\frac{\underset{\sim}{q}}{\theta}\cdot d\underset{\sim}{a} - \int_{V-\sigma}\frac{\rho h}{\theta}\, dv \geq 0 . \tag{2.5.17}$$

For adiabatic processes, $\underset{\sim}{q} = \underset{\sim}{0}$, $h = 0$, and this gives

$$\frac{D}{Dt}\int_{V-\sigma}\rho\eta\, dv \geq 0 \tag{2.5.18}$$

which is the statement that <u>in an adiabatic process the total entropy cannot decrease</u>.

The entropy inequalities (2.5.15) and (2.5.16)

are the generalizations of the classical Clausius-Duhem inequalities.

Definition. The thermomechanical state of the body will be called thermodynamically admissible if and only if the entropy inequalities (2.5.15) and (2.5.16) are not violated.

As we shall see, thermodynamic admissibility plays a central role in the constitutive theory.

Chapter 3
CONSTITUTIVE THEORY

3.1. Scope of the Chapter

The fundamental axioms of the general theory of constitutive equations for nonlocal micropolar continuum theories are presented in Art. 3.2. The nonlinear constitutive equations are then obtained for anisotropic thermoelastic solids in Art. 3.3 and their special forms for isotropic materials are presented in Art. 3.4. The linear theories of anisotropic and isotropic micropolar thermoelasticity are discussed and the restrictions arising from the nonnegative character of the strain energy are obtained in Arts. 3.5 and 3.6.

3.2. Axioms of Constitutive Theory

The fundamental laws of the micropolar theory formulated in Chapter 2 consist of

Conservation of mass :

$$\frac{\partial \rho}{\partial t} + (\rho v^k)_{;k} = 0 \tag{3.2.1}$$

Conservation of microinertia :

$$\frac{\partial i^{\cdot kl}}{\partial t} + i^{\cdot kl}_{\;\;;m} v^m - i^{\cdot kr} \nu^{l}_{\;r} - i^{\cdot rl} \nu^{k}_{\;r} = 0 \tag{3.2.2}$$

Balance of momentum :

(3.2.3) $$t^{kl}{}_{;k} + \rho(f^l - \dot{v}^l) = 0$$

Balance of moment of momentum :

(3.2.4) $$m^{kl}{}_{;k} + \varepsilon^{lmn} t_{mn} + \rho[l^l - \overline{(j^{lm} \nu_m)}^{\,\cdot}] = 0$$

Conservation of energy :

(3.2.5) $$\rho\dot{\varepsilon} = t^{kl}(v_{l;k} + \nu_{kl}) + m^{kl}\nu_{l;k} + q^k{}_{;k} + \rho h$$

Entropy inequality :

(3.2.6) $$\rho\gamma \equiv \rho\dot{\eta} - \rho\theta^{-1}h - (q^k/\theta)_{;k} = 0 \ .$$

These consist of fourteen partial differential equations and one inequality. Given f^l, l^l and h there are thirty-seven unknowns $\rho, v^k, \dot{i}^{kl}, \nu^k, t^{kl}, m^{kl}, q^k, \varepsilon, \eta$ and θ. Clearly the system (3.2.1) to (3.2.6) is highly indeterminate. Twenty-three additional independent equations are needed for the determination of the motion and deformation of a micropolar body. This is clear however from another observation, namely, equations (3.2.1) to (3.2.6) are valid for any micropolar body irrespective of its physical constitution, e.g., whether it is solid, fluid, gaseous, plastic, etc. The material constitution must be considered in order to restrict these equations further. This can be done from a molecular or continuum viewpoint. Statistical mechanical methods are not yet available for the exact treatment of this problem for the general class of bodies. Although special theories are constructed for dilute gases (Maxwell-Boltzmann theory of dilute gases) and for elastic solids (Born-Kármán theory of lattice dynamics), these theories deal with special types of media.

In the spirit of continuum mechanics, a set of general axioms has evolved during the past decade. Our extensions and organizations of these axioms, Eringen [1966 a] , [1967 d] , will be applied to micropolar thermoelastic solids. These axioms are :

(I) axiom of Causality

(II) axiom of Determinism

(III) axiom of Equipresence

(IV) axiom of Objectivity

(V) axiom of Material Invariance

(VI) axiom of Neighborhood

(VII) axiom of Memory

(VIII) axiom of Admissibility

(I) Axiom of Causality : The motion of material particles of a body and their temperatures are considered to be self-evident observable effects in every thermomechanical behavior of the body. The remaining quantities (other than those derivable from the temperature and motion), excluding the prescribed source terms that enter the balance laws and entropy inequality, are "causes" or dependent constitutive variables.

Thus the independent constitutive variables are

$$(3.2.7) \qquad x^k = x^k(\underset{\sim}{X}, t) \;,\; x^k_{\;K} = x^k_{\;K}(\underset{\sim}{X}, t) \;,\; \theta(X, t) .$$

The dependent constitutive variables are

$$(3.2.8) \qquad t^{kl} \;,\; m^{kl} \;,\; q^k \;,\; \varepsilon \;,\; \eta \;.$$

Note that ρ, v^k, i^{kl} and ν^k are derivable from (3.2.7) through either basic laws or the kinematical relations.

(II) Axiom of Determinism : The value of thermo-mechanical constitutive functions at a material particle $\underset{\sim}{X}$ of the body $\mathcal{B}$ at time t is determined by the history of the motions and temperatures of all material particles of $\mathcal{B}$. Mathematically

$$\underset{\sim}{t}(\underset{\sim}{X}, t) = \underset{\sim}{F}[\underset{\sim}{x}(\underset{\sim}{X}', t'), \underset{\sim}{x}_K(X', t'), \theta(\underset{\sim}{X}', t'), \underset{\sim}{X}, t]$$

$$\underset{\sim}{m}(\underset{\sim}{X}, t) = \underset{\sim}{M}[\underset{\sim}{x}(\underset{\sim}{X}', t'), \underset{\sim}{x}_K(X', t'), \theta(\underset{\sim}{X}', t'), \underset{\sim}{X}, t]$$

$$(3.2.9) \qquad \underset{\sim}{q}(\underset{\sim}{X}, t) = \underset{\sim}{G}[\underset{\sim}{x}(\underset{\sim}{X}', t'), \underset{\sim}{x}_K(X', t'), \theta(\underset{\sim}{X}', t'), \underset{\sim}{X}, t]$$

$$\varepsilon(\underset{\sim}{X}, t) = E[\underset{\sim}{x}(\underset{\sim}{X}', t'), \underset{\sim}{x}_K(X', t'), \theta(\underset{\sim}{X}', t'), \underset{\sim}{X}, t]$$

$$\eta(\underset{\sim}{X}, t) = N[\underset{\sim}{x}(\underset{\sim}{X}', t'), \underset{\sim}{x}_K(X', t'), \theta(\underset{\sim}{X}', t'), \underset{\sim}{X}, t]$$

where $\underset{\sim}{F}$ and $\underset{\sim}{M}$ are tensor-valued, $\underset{\sim}{G}$ is vector-valued, and ε and η are scalar-valued functionals over the fields of real functions $\underset{\sim}{x}(\underset{\sim}{X}', t')$, $\underset{\sim}{x}_K(\underset{\sim}{X}', t')$ and $\theta(\underset{\sim}{X}', t')$ for all points $\underset{\sim}{X}'$ of B at all times $t' \leq t$. They are also functions of $\underset{\sim}{X}$ and t. The prime is used to indicate these functional dependences.

(III) Axiom of Equipresence : At the outset, all constitutive functionals shall be dependent on the same list of independent constitutive variables. This is a "bookkeeping" system not allowing prejudice against any one set. Expressions (3.2.9) contain the same list of constitutive independent variables (3.2.7).

(IV) Axiom of Objectivity : The constitutive equations must be form invariant with respect to rigid macromotions of the spatial frame of reference and the constant time shift.

If two macromotions $\underset{\sim}{x}(\underset{\sim}{X}, t)$ and $\underset{\sim}{\bar{x}}(\underset{\sim}{X}, t)$ differ by a rigid motion and constant time shift, they are related to each other by

$$\underset{\sim}{\bar{x}}(\underset{\sim}{X}, \bar{t}) = \underset{\sim}{Q}(t)\underset{\sim}{x}(\underset{\sim}{X}, t) + \underset{\sim}{b}(t) \tag{3.2.10}$$

where $\underset{\sim}{Q}(t)$ is a proper orthogonal transformation, $\underset{\sim}{b}(t)$ is a translation and $\bar{t}$ is shifted time, i.e.,

$$\begin{gathered} \underset{\sim}{Q}\underset{\sim}{Q}^T = \underset{\sim}{Q}^T\underset{\sim}{Q} = \underset{\sim}{I}, \det \underset{\sim}{Q} = +1 \\ \bar{t} = t - a \end{gathered} \tag{3.2.11}$$

where $\underset{\sim}{I}$ is the unit tensor and a is a constant.

The constitutive equations are form invariant under the transformations (3.2.10) for all members of the proper orthogonal group $\{\underset{\sim}{Q}(t)\}$, for all translations $\{\underset{\sim}{b}(t)\}$ and time shifts a. This is stated as : constitutive functionals are hemitropic spatially. If one includes reflection by considering the full group for $\underset{\sim}{Q}$ (in which det $Q = \pm 1$), then the constitutive functionals are said to be spatially isotropic.

Under the transformations (3.2.10), ε and η remain invariant and $\underset{\sim}{t}$, $\underset{\sim}{m}$ and $\underset{\sim}{q}$ transform as

(3.2.12)
$$\begin{aligned}\bar{\underset{\sim}{t}}(\underset{\sim}{X}, \bar{t}) &= \underset{\sim}{Q}(t)\underset{\sim}{t}(\underset{\sim}{X}, t)\underset{\sim}{Q}^T(t)\\ \bar{\underset{\sim}{m}}(\underset{\sim}{X}, \bar{t}) &= \underset{\sim}{Q}(t)\underset{\sim}{m}(\underset{\sim}{X}, t)\underset{\sim}{Q}^T(t)\\ \bar{\underset{\sim}{q}}(\underset{\sim}{X}, \bar{t}) &= \underset{\sim}{Q}(t)\underset{\sim}{q}(\underset{\sim}{X}, t)\ .\end{aligned}$$

The transformation of the argument functions of (3.2.9) are

(3.2.13)
$$\begin{aligned}\bar{\underset{\sim}{x}}(\underset{\sim}{X}', \bar{t}') &= \underset{\sim}{Q}(t')\underset{\sim}{x}(\underset{\sim}{X}', t') + \underset{\sim}{b}(t')\\ \bar{\underset{\sim}{\chi}}_K(\underset{\sim}{X}', \bar{t}') &= \underset{\sim}{Q}(t')\underset{\sim}{\chi}_K(\underset{\sim}{X}', t')\end{aligned}$$

so that objectivity (alternatively material frame indifference) for the stress constitutive equation is satisfied if and only if

(3.2.14)
$$\begin{aligned}&\underset{\sim}{Q}(t)\underset{\sim}{F}[\underset{\sim}{x}(\underset{\sim}{X}', t'), \underset{\sim}{\chi}_K(\underset{\sim}{X}', t'), \theta(\underset{\sim}{X}', t'), \underset{\sim}{X}, t]\underset{\sim}{Q}^T(t) = \underset{\sim}{F}[\underset{\sim}{Q}(t')\cdot\\ &\cdot\underset{\sim}{x}(\underset{\sim}{X}', t') + \underset{\sim}{b}(t'), \underset{\sim}{Q}(t)\underset{\sim}{\chi}_K(\underset{\sim}{X}', t'), \theta(\underset{\sim}{X}', t'), \underset{\sim}{X}, \bar{t}]\ .\end{aligned}$$

Similar conditions are written for $\underset{\sim}{m}$, $\underset{\sim}{q}$, ε and η.

The following three special transformations are completely equivalent to satisfying the requirement of objectivity :

(a) <u>The rigid Translation of the Spatial Frame</u>. In this case, $\underset{\sim}{Q}(t') = \underset{\sim}{I}$, $a = 0$, $\underset{\sim}{b}(t') = -\underset{\sim}{x}(\underset{\sim}{X}, t')$. This means translation of the origin of the spatial frame of reference to $\underset{\sim}{x}$. From (3.2.10) and (3.2.11) we have

$$\bar{x}(\underset{\sim}{X}', \bar{t}') = \underset{\sim}{x}(\underset{\sim}{X}', t') - \underset{\sim}{x}(\underset{\sim}{X}, t'), \quad \bar{t}' = t' \qquad (3.2.15)$$

(b) <u>Shift of time</u>. Selecting $\underset{\sim}{Q}(t') = \underset{\sim}{I}$, $\underset{\sim}{b}(t') = 0$, $a = t$, we have

$$\bar{\underset{\sim}{x}}(\underset{\sim}{X}', \bar{t}') = x(\underset{\sim}{X}', t'), \quad \bar{t}' = t' - t, \quad -\infty \leq \bar{t}' \leq 0 .$$

If we also write

$$\tau' \equiv t - t' \geq 0 \qquad 0 \leq \tau' \leq \infty \qquad (3.2.16)$$

and use (3.2.15), equations (3.2.14) reduce to

$$\underset{\sim}{t}(\underset{\sim}{X}, t) = \underset{\sim}{F}[\underset{\sim}{x}(\underset{\sim}{X}', t - \tau') - \underset{\sim}{x}(\underset{\sim}{X}, t - \tau'), \underset{\sim}{x}_{K}(\underset{\sim}{X}', t - \tau'), \theta(\underset{\sim}{X}', t - \tau'), \underset{\sim}{X}] \qquad (3.2.17)$$

(c) <u>Rigid Rotations of the Spatial Frame</u>. We take $\underset{\sim}{b} = \underset{\sim}{0}$, $a = 0$, and $\underset{\sim}{Q}(t')$ arbitrary. Since $\underset{\sim}{t}$ transforms as in (3.2.14), applying (3.2.17) gives

$$\begin{aligned} &\underset{\sim}{Q}(t)\underset{\sim}{F}[\underset{\sim}{x}(\underset{\sim}{X}', t - \tau') - \underset{\sim}{x}(\underset{\sim}{X}, t - \tau'), \underset{\sim}{x}_{K}(\underset{\sim}{X}', t - \tau'), \theta(\underset{\sim}{X}', t - \tau'), \underset{\sim}{X}]\underset{\sim}{Q}^{T}(t) \\ &= \underset{\sim}{F}\{\underset{\sim}{Q}(t - \tau')[\underset{\sim}{x}(\underset{\sim}{X}', t - \tau') - \underset{\sim}{x}(\underset{\sim}{X}, t - \tau'), \underset{\sim}{Q}(t - \tau')\underset{\sim}{x}_{K}(\underset{\sim}{X}', t - \tau'), \theta(\underset{\sim}{X}', t - \tau'), \underset{\sim}{X}\}. \end{aligned} \qquad (3.2.18)$$

These restrictions are valid for every member of the group $\{Q\}$ for $\tau' \geq 0$, subject to (3.2.11).

Equations (3.2.17), subject to the restrictions

(3.2.18), are the most general constitutive equations for non-simple materials with memory. Since we are interested in only micropolar thermoelasticity (i.e., no memory), the functionals are independent of the past time. Formally this is achieved by taking $\tau' \equiv 0$ in (3.2.17) and (3.2.18).

(V) Axiom of Material Invariance : The materials, generally, possess some symmetry conditions in their material state. This may be characterized by a group of orthogonal transformations $\{\underset{\sim}{S}\}$ and translations $\{\underset{\sim}{B}\}$ of the material frame of reference $\underset{\sim}{X}$. If the constitutive equations (3.2.9) are form invariant under the transformations

$$(3.2.19) \qquad \bar{\underset{\sim}{X}} = \underset{\sim}{S}\underset{\sim}{X} + \underset{\sim}{B} \;, \quad \underset{\sim}{S}\underset{\sim}{S}^T = \underset{\sim}{S}^T\underset{\sim}{S} = \underset{\sim}{I} \;, \quad \det \underset{\sim}{S} = \pm 1$$

then the material is said to possess the symmetry characterized by the groups $\{\underset{\sim}{S}\}$ and $\{\underset{\sim}{B}\}$. For the isotropic materials, for example, $\{\underset{\sim}{S}\}$ is the full orthogonal group, and for the homogeneous materials the constitutive equations are not explicit functions of $\underset{\sim}{X}$. In the classical theory of elasticity, all thirty-two members of the crystal classes can be characterized by twelve members of the full orthogonal group $\{\underset{\sim}{S}\}$.

(VI) Axiom of Neighborhood : The values of the independent constitutive variables at distant material points from $\underset{\sim}{X}$ do not affect appreciably the values of the constitutive dependent variables at $\underset{\sim}{X}$. This provides an intuitive guidance for approximation. Mathematically it is an expression of the continuity requirements on the constitutive functionals. Physically it is the realization of the fact that intermolecular and binding forces are of short range and the atoms and molecules are not greatly affected by the forces of their distant neighbors. The axiom of neighborhood is a statement of <u>locality</u>. The general mathematical statement of the axiom of neighborhood was formulated by Eringen [1966 a] . Here we proceed to give the result for the case called <u>smooth neighborhood.</u>

Assuming that the motions possess Taylor series expansions (in cartesian coordinates)

$$\begin{aligned}
\underset{\sim}{x}(\underset{\sim}{X}', t-\tau') &= \underset{\sim}{x}(\underset{\sim}{X}, t-\tau') + (X'^K - X^K)\, x_{,K}(\underset{\sim}{X}, t-\tau') \\
&\quad + \frac{1}{2!}(X'^K - X^K)(X'^L - X^L)\,\underset{\sim}{x}_{,KL}(\underset{\sim}{X}, t-\tau') + \dots \\
\underset{\sim}{\chi}_M(X', t-\tau') &= \underset{\sim}{\chi}_M(\underset{\sim}{X}, t-\tau') + (X'^K - X^K)\,\underset{\sim}{\chi}_{M,K}(\underset{\sim}{X}, - t-\tau') + \dots \\
\theta(\underset{\sim}{X}', t-\tau') &= \theta(\underset{\sim}{X}, t-\tau') + (X'^K - X^K)\,\theta_{,K}(\underset{\sim}{X}, t-\tau') + \dots
\end{aligned} \tag{3.2.20}$$

Now if the constitutive functionals are smooth enough so that they can be expressed as functionals in the new functions

$$\begin{aligned}
&\underset{\sim}{x}_{,K}(t-\tau') \;, \quad \underset{\sim}{x}_{,KL}(t-\tau'), \dots \\
&\underset{\sim}{\chi}_K(t-\tau') \;, \quad \underset{\sim}{\chi}_{K,L}(t-\tau'), \dots \\
&\theta(t-\tau') \;, \quad \theta_{,K}(t-\tau'), \dots
\end{aligned} \tag{3.2.21}$$

then we say that the material possesses <u>smooth neighborhood</u>. In the list (3.2.21), we left out the argument $\underset{\sim}{X}$, since the functionals are no longer of the space-type. Thus for such materials, (3.2.17) is replaced by

$$\underset{\sim}{t}(\underset{\sim}{X},t) = \underset{\sim}{F}[\underset{\sim}{x}_{,K}(t-\tau'), \underset{\sim}{x}_{,KL}(t-\tau'), \ldots; \underset{\sim}{\chi}_{K}(t-\tau'), \underset{\sim}{\chi}_{K,L}(t-\tau'), \ldots,$$
(3.2.22)
$$\theta(t-\tau'),\ \theta_{,K}(t-\tau'), \ldots;\ \underset{\sim}{X}, \underset{\sim}{D}'_{K}]$$

where $\underset{\sim}{D}'_{K}$ are three material directors arising from the fact that dependence of $\underset{\sim}{X}'-\underset{\sim}{X}$ at $\underset{\sim}{X}$ is equivalent to dependence on three vectors.For local theories,we may select $\underset{\sim}{D}_{K}=\underset{\sim}{G}_{K}$ .In this case,no need for explicit statement of this dependence is necessary,since it naturally arises from the material gradients.

<u>Definition</u>. <u>Micropolar thermoelastic solids with memory are characterized by the constitutive equations of the form</u>

$$\underset{\sim}{t}(\underset{\sim}{X},t) = \underset{\sim}{F}[\underset{\sim}{x}_{,K}(t-\tau'), \underset{\sim}{\chi}_{K}(t-\tau'), \underset{\sim}{\chi}_{K,L}(t-\tau'), \theta(t-\tau'), \theta_{,K}(t-\tau'), \underset{\sim}{X}].$$
(3.2.23)

If these materials possess no memory,then the constitutive functionals reduce to ordinary functions. Thus <u>micropolar thermoelastic materials possess constitutive equations</u> of the form

$$\underset{\sim}{t}(\underset{\sim}{X},t) = \underset{\sim}{F}(\underset{\sim}{x}_{,K}, \underset{\sim}{\chi}_{K}, \underset{\sim}{\chi}_{K,L}, \theta, \theta_{,K}, \underset{\sim}{X})$$

$$\underset{\sim}{m}(\underset{\sim}{X},t) = \underset{\sim}{M}(\underset{\sim}{x}_{,K}, \underset{\sim}{\chi}_{K}, \underset{\sim}{\chi}_{K,L}, \theta, \theta_{,K}, \underset{\sim}{X})$$

(3.2.24a) $$\underset{\sim}{q}(\underset{\sim}{X},t) = \underset{\sim}{Q}(\underset{\sim}{x}_{,K}, \underset{\sim}{\chi}_{K}, \underset{\sim}{\chi}_{K,L}, \theta, \theta_{,K}, \underset{\sim}{X})$$

$$\varepsilon(\underset{\sim}{X}, t) = E(\underset{\sim}{x}_{,K}, \underset{\sim}{\chi}_K, \underset{\sim}{\chi}_{K,L}, \theta, \theta_{,K}, \underset{\sim}{X})$$

$$\eta(\underset{\sim}{X}, t) = N(\underset{\sim}{x}_{,K}, \underset{\sim}{\chi}_K, \underset{\sim}{\chi}_{K,L}, \theta, \theta_{,K}, \underset{\sim}{X}) \qquad (3.2.24b)$$

subject to the axiom of objectivity of the form (3.2.18). Thus the micropolar thermoelastic solids generalize the classical elastic solids by having a dependence on the motion of the directors and their gradients.

(VII) Axiom of Memory : The values of the constitutive variables at a distant past do not affect appreciably the values of the constitutive functionals at the present. The micropolar elastic materials are independent of memory, and therefore we need no reference to this axiom. In connection with its uses, we refer the reader to Colemann & Noll [1961] , Truesdell and Noll [1965] , and Eringen [1962] , [1967 d] .

(VIII) Axiom of Admissibility : All constitutive equations must be consistent with the basic principles of continuum mechanics, that is they are subject to balance laws and the entropy inequality. This seemingly obvious statement has many important consequences. For example, since through the conservation of mass

$$\frac{\rho_0(\underset{\sim}{X}')}{\rho(\underset{\sim}{X}', t')} = \det \frac{\partial x^k(\underset{\sim}{X}', t')}{\partial X'^K}$$

we see that dependence on ρ is implied by the dependence on $\partial x^k / \partial X'^K$. In certain cases (e.g., fluids), the constitutive functionals may depend on det $(\partial x^k/\partial X'^K)$. This implies the dependence on the density ρ.

Another important consequence of this axiom is that constitutive functionals must obey the Clausius-Duhem inequality. When this inequality is assumed to apply for all independent thermomechanical processes, the consequence is often very dramatic, as will be demonstrated in the following section.

3.3. Constitutive Equations of Micropolar Thermoelastic Solids

Definition. A body is called a micropolar thermoelastic solid if the stress, couple stress, heat, free energy, and entropy have the forms :

$$
\begin{aligned}
t^{kl}(\underset{\sim}{X}, t) &= t^{kl}(x^k_{,K}, \varphi^k, \varphi^k_{:K}, \theta, \theta_{,K}, \underset{\sim}{X}) \\
m^{kl}(\underset{\sim}{X}, t) &= m^{kl}(x^k_{,K}, \varphi^k, \varphi^k_{:K}, \theta, \theta_{,K}, \underset{\sim}{X}) \\
q^k(\underset{\sim}{X}, t) &= q^k(x^k_{,K}, \varphi^k, \varphi^k_{:K}, \theta, \theta_{,K}, \underset{\sim}{X}) \qquad (3.3.1) \\
\psi(\underset{\sim}{X}, t) &= \psi(x^k_{,K}, \varphi^k, \varphi^k_{:K}, \theta, \theta_{,K}, \underset{\sim}{X}) \\
\eta(\underset{\sim}{X}, t) &= \eta(x^k_{,K}, \varphi^k, \varphi^k_{:K}, \theta, \theta_{,K}, \underset{\sim}{X}) .
\end{aligned}
$$

Note that for micropolar bodies, $\underset{\sim}{\chi}$ is an orthogonal tensor, so that $\underset{\sim}{\chi}_K$ and $\underset{\sim}{\chi}_{K;L}$ appearing in (3.2.24) are replaced by $\underset{\sim}{\varphi}$ and $\underset{\sim}{\varphi}_{,L}$. Further since $\psi = \varepsilon - \theta\eta$, (3.2.24) implies the dependence

$(3.3.1)_4$ for the free energy. Thus the set (3.3.1) is equivalent to (3.2.24).

To obtain the consequence of the Clausius-Duhem inequality (2.5.15), we carry (3.3.1) into (2.5.15). First we note that

$$\dot{\psi} = \frac{\partial \psi}{\partial x^k} v^k{}_{;\ell} x^\ell{}_{,K} + \frac{\partial \psi}{\partial \varphi^k} \dot{\varphi}^k + \frac{\partial \psi}{\partial \varphi^k{}_{:K}} \dot{\varphi}^k{}_{:K} + \frac{\partial \psi}{\partial \theta} \dot{\theta} + \frac{\partial \psi}{\partial \theta_{,K}} \dot{\theta}_{,K}$$

where we used (1.6.10). Thus we have

$$-\frac{\rho}{\theta}\left(\frac{\partial \psi}{\partial \theta} + \eta\right)\dot{\theta} - \frac{\rho}{\theta}\frac{\partial \psi}{\partial \theta_{,K}}\dot{\theta}_{,K} + \frac{1}{\theta}\left(t^k{}_\ell - \rho x^k{}_{,K}\frac{\partial \psi}{\partial x^\ell{}_{,K}}\right) v^\ell{}_{;k} + \frac{1}{\theta}\left(m^\ell{}_k X^K{}_{,\ell} \Lambda^k{}_n - \right.$$

$$\left. -\rho \frac{\partial \psi}{\partial \varphi^n{}_{:K}}\right) \dot{\varphi}^n{}_{:K} + \frac{1}{\theta}\left(m^\ell{}_k X^K{}_{,\ell} \Lambda^k{}_{n:K} - \varepsilon_{mk\ell} \Lambda^m{}_n t^{k\ell} - \rho \frac{\partial \psi}{\partial \varphi^n}\right)\dot{\varphi}^n +$$

$$+ \frac{q^k}{\theta^2}\theta_{,k} \geq 0 \quad \text{in } V - \sigma$$

which must be satisfied for all independent variations of $\dot{\theta}$, $v^k{}_{;\ell}$, $\dot{\varphi}^n$, $\dot{\varphi}^n{}_{:K}$, $\dot{\theta}_{,K}$ and $\theta_{,k}$. Since this equation is linear in the first five of these quantities, it cannot be maintained unless the coefficients of these quantities vanish. Thus

$$\eta = -\frac{\partial \psi}{\partial \theta} \quad , \quad \frac{\partial \psi}{\partial \theta_{,K}} = 0$$

$$t^k{}_\ell = \rho x^k{}_{,K} \cdot \frac{\partial \psi}{\partial x^\ell{}_{,K}} \tag{3.3.2a}$$

$$m^{\ell}{}_{k} = \rho x^{\ell}{}_{,K} \overset{-1}{\Lambda}{}^{n}{}_{\ell} \frac{\partial \psi}{\partial \varphi^{n}{}_{:K}}$$

$$0 = \rho \frac{\partial \psi}{\partial \varphi^{n}} + \varepsilon_{rk\ell} t^{k\ell} \Lambda^{r}{}_{n} - m^{\ell}{}_{k} \Lambda^{k}{}_{n:\ell}$$

(3.3.2b) $$\frac{1}{\theta^2} q^{k} \theta_{,k} \geq 0 \quad \text{in} \quad V - \sigma \ .$$

This last equation must be valid for all independent processes.

An interestint consequence of $(3.3.2)_5$ is that it contains the micropolar statement of the axiom of objectivity. In fact, multiplying this by $\overset{-1}{\Lambda}{}^{n}{}_{m}$ and using $(3.3.2)_3$, $(3.3.2)_4$, and the fact that $\underset{\sim}{\Lambda}\underset{\sim}{\Lambda}^{-1} = \underset{\sim}{I}$, we obtain

(3.3.3) $$\varepsilon_{m\ell}{}^{k} x^{\ell}{}_{,L} \frac{\partial \psi}{\partial x^{k}{}_{,L}} + \overset{-1}{\Lambda}{}^{\ell}{}_{m} \frac{\partial \psi}{\partial \varphi^{\ell}} + \overset{-1}{\Lambda}{}^{\ell}{}_{m:L} \frac{\partial \psi}{\partial \varphi^{\ell}{}_{:L}} = 0 \ .$$

We observe that for $\partial \psi / \partial \varphi^{\ell} = 0$, $\partial \psi / \partial \varphi^{\ell}{}_{:L} = 0$ these differential equations possess the general solution

(3.3.4) $$\psi = \psi(C_{KL}) \ , \quad C_{KL} \equiv g_{k\ell} x^{k}{}_{,K} x^{\ell}{}_{,L} \ .$$

Thus the general solution of (3.3.3) must lead to the most general form of ψ which satisfies the axiom of objectivity. In fact, the general solution of the system (3.3.3) is found to be (Kafadar and Eringen [1970])

(3.3.5) $$\psi = \psi(\mathcal{C}_{KL}, \Gamma_{KL}, \theta, \underset{\sim}{X})$$

where

$$\mathcal{C}_{KL} \equiv x^k_{,K} \chi_{kL}$$
$$\Gamma_{KL} \equiv \frac{1}{2} \varepsilon_{KMN} \chi^{kM}{}_{:L} \chi_k{}^N \tag{3.3.6}$$

are the deformation tensors introduced before.

From the heat vector we form the three spatial scalars

$$Q^K = q^k X^K_{,k} \ .$$

These scalars are functions of the same list of variables contained in q^k, since $X^K_{,k}$, is a function of $x^k_{,K}$ (cf. (1.4.15)). The minimum function basis for a scalar function of the twenty-one independent quantities $x^k_{,K}$, φ^k, $\varphi^k_{:K}$ is 21 - 3 = 18. Since $\mathcal{C}_{KL}$ and Γ_{KL} constitutive such a basis, Q^K must be expressible as functions of these eighteen variables. This follows from Cauchy's theorem on scalar invariants of vectors (cf. Eringen, Appendix B). Solving the above equation, we obtain

$$q^k = Q^K(\underset{\sim}{\mathcal{C}}, \underset{\sim}{\Gamma}, \theta, \theta_{,K}, \underset{\sim}{X}) x^k_{,K} \ .$$

Substituting (3.3.5) into (3.3.2), we have for the complete set of constitutive equations :

$$t^{kl} = \rho \frac{\partial \psi}{\partial \mathcal{C}_{KL}} x^k_{,K} \chi^l_{\ L}$$
$$m^{kl} = \rho \frac{\partial \psi}{\partial \Gamma_{LK}} x^k_{,K} \chi^l_{\ L} \tag{3.3.7a}$$

$$q^k = Q^K(\underset{\sim}{\mathcal{C}}, \underset{\sim}{\Gamma}, \theta, \underset{\sim}{\nabla}\theta, \underset{\sim}{X})\, x^k_{,K}$$

$$\psi = \psi(\underset{\sim}{\mathcal{C}}, \underset{\sim}{\Gamma}, \theta, \underset{\sim}{X})$$

(3.3.7b) $$\eta = -\frac{\partial \psi}{\partial \theta}$$

subject to

(3.3.8) $$Q^K \theta_{,K} \geq 0$$

for all independent processes. Equations (3.3.7) are the <u>non-linear constitutive equations of inhomogeneous, anisotropic, micropolar thermoelastic solids</u>.

The above analysis imposed by the axiom of thermodynamics admissibility indicates how this innocent axiom leads to simple concrete results. The great simplification is in the reduction of the number of constitutive functions from twenty-three to four, by finding a scalar function ψ from which all constitutive functions (except $\underset{\sim}{q}$) are derivable. In addition, we have the extra benefit in having satisfied the axiom of objectivity for these constitutive functions. The progress however stops at the heat vector. In fact, for all dissipative processes we face such difficulties. Nevertheless, still this method provides the thermodynamic restrictions essential to material behavior and indicates directions for progress.

The inequality (3.3.8) implies one other important

result which depends on the continuity requirement of Q^K with respect to $\theta_{,K}$. If this is assumed, then we can show that Q^K vanishes with $\theta_{,K}$, i.e.,

$$Q^K(\underset{\sim}{\mathscr{C}}, \underset{\sim}{\Gamma}, \theta, \underset{\sim}{0}, \underset{\sim}{X}) = 0 \ . \tag{3.3.9}$$

To see this, suppose $Q^2 = Q^3 = 0$, $Q^1 \neq 0$ then

$$Q^1 \theta_{,1} \geq 0 \ .$$

But this implies that $Q^1 > 0$ whenever $\theta_{,1} > 1$ and $Q^1 < 0$ when $\theta_{,1} < 0$. If Q^1 is continuous in $\theta_{,1}$, then $Q^1 = 0$ when $\theta_{,1} = 0$. Similar arguments are valid for Q^2 and Q^3. Hence the proof of (3.3.9). It should be noted that instead of the deformation tensors $\mathscr{C}_{KL}$ and Γ_{KL}, other deformation measures can be used in the constitutive equations. For example, if $\overset{-1}{\underset{\sim}{\mathscr{C}}}$ is employed, the equivalent constitutive equation to $(3.3.7)_1$ is

$$t^k{}_\ell = -\rho \frac{\partial \psi}{\partial \overset{-1}{\mathscr{C}}{}^{KL}} \chi^{kK} X^L{}_{,\ell} \ . \tag{3.3.10}$$

For isotropic materials, still other deformation measures (spatial measures) produce simplifications.

<u>Incompressible solids</u> : A material is called incompressible if its mass density is constant. In this case

$$\frac{\rho_0}{\rho} = \det \mathscr{C}^K{}_L = III_{\mathscr{C}} = 1 \ . \tag{3.3.11}$$

This condition reduces the number of independent variables by one. Thus the partial derivatives of ψ with respect to $\underset{\sim}{\mathscr{C}}$ are not all independent. To remedy this situation, we introduce the Lagrange multiplier $-p/q_o$ by replacing ψ by

$$-\frac{p}{\rho_o}(III_{\mathcal{C}} - 1) + \psi \,.$$

Substituting this into (3.3.7) and noting that

$$\frac{\partial III_{\mathcal{C}}}{\partial \mathcal{C}_{MN}} = \frac{\rho_o}{\rho} \overset{-1}{\mathcal{C}}{}^{MN} \tag{3.3.12}$$

we obtain

$$t^{kl} = -\, p g^{kl} + \rho \frac{\partial \psi}{\partial \mathcal{C}_{KL}} x^k_{\ ,K} \chi^l_{\ L} \tag{3.3.13}$$

which must be supplemented with (3.3.11). Equations (3.3.13) are the stress constitutive equations for <u>incompressible</u> anisotropic micropolar solids. The unknown pressure $p(\underset{\sim}{X}, t)$ (not to be confused with the thermodynamic pressure) must be determined upon integration of the differential equations of motion and use of the boundary conditions.

The <u>axiom of material invariance</u> imposes further restrictions on the constitutive equations. If the material in the natural state possesses restrictions characterized by a group of rotations $\{\underset{\sim}{S}\}$ and translations $\{\underset{\sim}{B}\}$ of the material frame of reference, then the constitutive equations must be form invariant under the transformations

$$\overline{X} = \underset{\sim}{S}\underset{\sim}{X} + \underset{\sim}{B} \tag{3.3.14}$$

$$\underset{\sim}{S}\underset{\sim}{S}^T = \underset{\sim}{S}^T\underset{\sim}{S} = \underset{\sim}{I}, \qquad \det \underset{\sim}{S} = \pm 1 \,.$$

According to the axiom of material invariance, this means that

$$\psi(\bar{\mathfrak{C}}, \bar{\Gamma}, \theta, \bar{X}) = \psi(\mathfrak{C}, \Gamma, \theta, X)$$

for all members of $\{S\}$ and $\{B\}$ where

$$\bar{\mathfrak{C}} = S\mathfrak{C}S^T \ , \quad \bar{\Gamma} = S\Gamma S^T \ .$$

Thus

$$\psi(S\mathfrak{C}S^T, S\Gamma S^T, \theta, SX+B) = \psi(\mathfrak{C}, \Gamma, \theta, X) \ . \qquad (3.3.15)$$

If the material is homogeneous, then dependence on B cannot occur, for by taking $S = I, B = -X$ we find that ψ is independent of X. If the material is isotropic then S is the full group. Thirty-two different classes of crystalline solids are obtained by the use of eleven members of the group $\{S\}$ and a reflection of the axes of X. In the next section, we discuss only the case of isotropic micropolar solids.

3.4. Isotropic Micropolar Solids

A micropolar elastic solid is isotropic if and only if the symmetry group S is the full group. In this case, the axiom of material invariance expressed by (3.3.15) implies that Ψ will be a function of invariants of $\mathfrak{C}$ and Γ. The total number of independent joint invariants of these tensors is fifteen. From (1.4.28) and (1.4.29) equivalently we may consider ψ a function of θ and the joint invariants of $\mathfrak{E}, \Upsilon$

or $\overset{-1}{\underset{\sim}{C}}$, $\underset{\sim}{\gamma}$. In these cases

(3.4.1)
$$t^{k\ell} = \rho \frac{\partial \psi}{\partial C_{\ell m}} C^{k}{}_{m}$$

$$m^{k\ell} = \rho \frac{\partial \psi}{\partial \gamma_{\ell m}} C^{k}{}_{m}$$

where

$$\psi = \psi(\underset{\sim}{C}, \underset{\sim}{\gamma}, \theta)$$

and

(3.4.2)
$$t^{k\ell} = -\rho \frac{\partial \psi}{\partial \overset{-1}{C}_{mk}} \overset{-1}{C}_{m}{}^{\ell}$$

$$m^{k\ell} = \rho \frac{\partial \psi}{\partial \gamma_{\ell m}} C^{k}{}_{m}$$

where

$$\psi = \psi(\overset{-1}{\underset{\sim}{C}}, \underset{\sim}{\gamma}, \theta) .$$

The free energy ψ may be taken as a <u>function</u> of the following fifteen invariants, (On the question of single-valuedness see Kafadar and Eringen [1970], App. A).

$$I_1 \equiv \operatorname{tr} \underset{\sim}{C} , \quad I_2 \equiv \frac{1}{2} \operatorname{tr} \underset{\sim}{C}^2 , \quad I_3 \equiv \frac{1}{3} \operatorname{tr} \underset{\sim}{C}^3 ,$$

$$I_4 \equiv \frac{1}{2} \operatorname{tr} \underset{\sim}{C} \underset{\sim}{C}^T , \quad I_5 \equiv \operatorname{tr} \underset{\sim}{C}^2 \underset{\sim}{C}^T , \quad I_6 \equiv \frac{1}{2} \operatorname{tr} \underset{\sim}{C}^2 \underset{\sim}{C}^T \underset{\sim}{C}^T$$

(3.4.3a)
$$I_7 = \operatorname{tr} \underset{\sim}{C} \underset{\sim}{\gamma} , \quad I_8 \equiv \operatorname{tr} \underset{\sim}{C} \underset{\sim}{\gamma}^2 , \quad I_9 \equiv \operatorname{tr} \underset{\sim}{C}^2 \underset{\sim}{\gamma}$$

$$I_{10} \equiv \operatorname{tr} \boldsymbol{\gamma} \ , \quad I_{11} = \frac{1}{2} \operatorname{tr} \boldsymbol{\gamma}^2 , \quad I_{12} = \frac{1}{3} \operatorname{tr} \boldsymbol{\gamma}^3$$

$$I_{13} = \frac{1}{2} \operatorname{tr} \boldsymbol{\gamma} \boldsymbol{\gamma}^T \ , \quad I_{14} = \operatorname{tr} \boldsymbol{\gamma}^2 \boldsymbol{\gamma}^T , \quad I_{15} = \frac{1}{2} \operatorname{tr} \boldsymbol{\gamma}^2 \boldsymbol{\gamma}^T \boldsymbol{\gamma}^T . \qquad (3.4.3b)$$

Substituting these into (3.4.1) gives

$$t^{k\ell} = C^k{}_m \sum_{\mu=1}^{15} \alpha_\mu \frac{\partial I_\mu}{\partial C_{\ell m}}$$

$$m^{k\ell} = C^k{}_m \sum_{\mu=1}^{15} \alpha_\mu \frac{\partial I_\mu}{\partial \gamma_{\ell m}} \qquad (3.4.4)$$

where

$$\alpha_\mu = \rho \frac{\partial \psi}{\partial I_\mu} \ . \qquad (3.4.5)$$

From (3.4.4) and (3.4.3), we have the nonlinear constitutive equations for micropolar elastic solids.

$$\mathbf{t} = \alpha_1 \mathbf{C} + \alpha_2 \mathbf{C}^2 + \alpha_3 \mathbf{C}^3 + \alpha_4 \mathbf{C} \mathbf{C}^T + \alpha_5 (\mathbf{C} \mathbf{C}^T \mathbf{C}^T + \mathbf{C}^2 \mathbf{C}^T + \mathbf{C} \mathbf{C}^T \mathbf{C})$$
$$+ \alpha_6 (\mathbf{C}^2 \mathbf{C}^T \mathbf{C}^T + \mathbf{C} \mathbf{C}^T \mathbf{C}^T \mathbf{C}) + \alpha_7 \mathbf{C} \boldsymbol{\gamma} + \alpha_8 \mathbf{C} \boldsymbol{\gamma}^2 + \alpha_9 (\mathbf{C}^2 \boldsymbol{\gamma} + \mathbf{C} \boldsymbol{\gamma} \mathbf{C}) \qquad (3.4.6)$$

$$\mathbf{m} = \alpha_7 \mathbf{C}^2 + \alpha_8 (\mathbf{C}^2 \boldsymbol{\gamma} + \mathbf{C} \boldsymbol{\gamma} \mathbf{C}) + \alpha_9 \mathbf{C}^3 + \alpha_{10} \mathbf{C} + \alpha_{11} \mathbf{C} \boldsymbol{\gamma} + \alpha_{12} \mathbf{C} \boldsymbol{\gamma}^2 + \alpha_{13} \mathbf{C} \boldsymbol{\gamma}^T$$
$$+ \alpha_{14} (\mathbf{C} \boldsymbol{\gamma}^T \boldsymbol{\gamma}^T + \mathbf{C} \boldsymbol{\gamma} \boldsymbol{\gamma}^T + \mathbf{C} \boldsymbol{\gamma}^T \boldsymbol{\gamma}) + \alpha_{15} (\mathbf{C} \boldsymbol{\gamma} \boldsymbol{\gamma}^T \boldsymbol{\gamma}^T + \mathbf{C} \boldsymbol{\gamma}^T \boldsymbol{\gamma}^T \boldsymbol{\gamma}) \ . \qquad (3.4.7)$$

Minor simplifications are possible by use of the Cayley-Hamilton theorem.

3.5. The Linear Theory

Various order approximate theories are deducible from the general constitutive equations (3.3.7) or from (3.4.6) and (3.4.7) for isotropic solids. Being now interested in the linear theory, we introduce the strain measure

$$\mathfrak{E}_{KL} = \mathcal{C}_{KL} - G_{KL} \tag{3.5.1}$$

in place of $\underset{\sim}{\mathcal{C}}$ and write

$$\Sigma(\underset{\sim}{\mathfrak{E}}, \underset{\sim}{\Gamma}, \theta, \underset{\sim}{X}) \equiv \rho_0 \psi . \tag{3.5.2}$$

The linear theory is an approximation based on the expansion of Σ into a quadratic polynomial

$$\rho_0 \psi = \Sigma_0 + \Sigma_1^{KL} \mathfrak{E}_{KL} + \frac{1}{2} \Sigma_2^{KLMN} \mathfrak{E}_{KL} \mathfrak{E}_{MN} + \Sigma_3^{KL} \Gamma_{KL} + \frac{1}{2} \Sigma_4^{KLMN} \Gamma_{KL} \Gamma_{MN} + \Sigma_5^{KLMN} \mathfrak{E}_{KL} \Gamma_{MN} \tag{3.5.3}$$

where Σ_1^{KL}, Σ_2^{KLMN}, ..., Σ_5^{KLMN} are functions of θ and $\underset{\sim}{X}$. The only symmetry restrictions on these coefficients are :

$$\Sigma_2^{KLMN} = \Sigma_2^{MNKL} , \quad \Sigma_4^{KLMN} = \Sigma_4^{MNKL} . \tag{3.5.4}$$

Substituting (3.5.3) into $(3.3.7)_1$ and $(3.3.7)_2$ we get

$$t^{kl} = \frac{\rho}{\rho_0} (\Sigma_1^{KL} + \Sigma_2^{KLMN} \mathfrak{E}_{MN} + \Sigma_5^{KLMN} \Gamma_{MN}) x^k_{,L} \chi^l_{K} \tag{3.5.5}$$

$$m^{kl} = \frac{\rho}{\rho_0} (\Sigma_3^{KL} + \Sigma_4^{KLMN} \Gamma_{MN} + \Gamma_5^{MNKL} \mathfrak{E}_{MN}) x^k_{,L} \chi^l_{K}$$

Further linearization requires recalling that

$$\frac{\rho}{\rho_o} \simeq 1 - I_{\tilde{\alpha}} \;, \quad x^k_{\;,K} = (G_{MK} + U_{M;K})\, g^{Mk}$$
$$\chi_L^{\;\ell} \simeq (G_{NL} + \Phi_{NL})\, g^{N\ell} = (G_{NL} - \varepsilon_{NLR}\Phi^R)\, g^{N\ell} \tag{3.5.6}$$

where

$$\Phi_{NL} \equiv -\varepsilon_{NLM}\,\Phi^M$$

and U_M is the displacement vector and Φ^R is the microrotation vector. In the spirit of the linear theory, we also employ the infinitesimal strain measures $\tilde{\mathfrak{E}}$, $\tilde{\Gamma}$

$$\tilde{\mathfrak{E}}_{KL} \equiv U_{L:K} - \varepsilon_{KLM}\,\Phi^M, \quad \tilde{\Gamma}_{KL} \equiv \Phi_{K:L} \;. \tag{3.5.7}$$

Substituting (3.5.6) and (3.5.7) into (3.5.5) and retaining only the linear terms in $\tilde{\mathfrak{E}}$ and $\tilde{\Gamma}$, we obtain

$$t^{k\ell} = (1 - I_{\tilde{\varepsilon}})\,\sigma_1^{\;k\ell} + \sigma_1^{\;m\ell}(\tilde{\varepsilon}_m^{\;k} - \varphi_m^{\;k}) + \sigma_1^{\;km}\varphi^\ell_{\;m} \tag{3.5.8}$$
$$+ \sigma_2^{\;k\ell mn}\,\tilde{\varepsilon}_{mn} + \sigma_5^{\;k\ell mn}\,\tilde{\gamma}_{mn}$$

$$m^{k\ell} = [(1 - I_{\tilde{\varepsilon}})\,\sigma_2^{\;k\ell} + \sigma_3^{\;\ell m}(\tilde{\varepsilon}_m^{\;k} - \varphi_m^{\;k}) + \sigma_3^{\;mk}\varphi^\ell_{\;m} \tag{3.5.9}$$
$$+ \sigma_4^{\;\ell kmn}\,\tilde{\gamma}_{mn} + \sigma_5^{\;mn\ell k}\,\tilde{\varepsilon}_{mn} \;.$$

The linearized (with respect to the strains and temperature gradient) constitutive equation for the heat vector is

$$q^k = \varkappa_1^{km} \theta_{,m} + \varkappa_2^{kmn} \tilde{\varepsilon}_{mn} + \varkappa_3^{kmn} \tilde{\gamma}_{mn} . \tag{3.5.10}$$

In (3.5.8) and (3.5.9), we wrote

$$\sigma_\alpha^{k\ell} \equiv \Sigma_\alpha^{KL} g_K^k g_L^\ell , \quad \sigma_\lambda^{k\ell mn} \equiv \Sigma_\lambda^{KLMN} g_K^k g_L^\ell g_M^m g_N^n , \qquad \begin{array}{l} \alpha = 1,3 \\ \lambda = 2,4,5 \end{array}$$

and, in the spirit of the linear theory, we set

$$\begin{aligned} \tilde{\varepsilon}_{k\ell} &\equiv u_{\ell;k} - \varepsilon_{k\ell m} \varphi^m \simeq \mathscr{E}_{KL} g^K_k g^L_\ell \\ \tilde{\gamma}_{k\ell} &= \varphi_{k;\ell} \simeq \Gamma_{KL} g^K_k g^L_\ell . \end{aligned} \tag{3.5.11}$$

In these constitutive equations, $\sigma_\alpha^{k\ell}$, $\sigma_\lambda^{k\ell mn}$, $\varkappa_1^{km}$, $\varkappa_2^{kmn}$ and $\varkappa_3^{kmn}$ are functions of θ and $\underset{\sim}{X}$.

Finally, q^k (3.5.10) must satisfy the Clausius-Duhem inequality (3.3.8) for all independent variations of $\theta_{,k}$, $\tilde{\varepsilon}_{k\ell}$, $\tilde{\gamma}_{k\ell}$, i.e.,

$$\varkappa_1^{km} \theta_{,k} \theta_{,m} + \varkappa_2^{kmn} \tilde{\varepsilon}_{mn} \theta_{,k} + \varkappa_3^{kmn} \tilde{\gamma}_{mn} \theta_{,k} \geq 0 .$$

This is satisfied if and only if

$$\varkappa_2^{kmn} = \varkappa_3^{kmn} = 0$$

and $\varkappa_1^{km}$ is nonnegative definite. Thus (3.5.10) becomes

$$q^k = \varkappa_1^{kl} \theta_{,l} \,. \qquad (3.5.12)$$

Equations (3.5.8), (3.5.9), and (3. 5.12) are the most general constitutive equations for an inhomogeneous anisotropic linear micropolar elastic solid. So far the linearization has been made with respect to temperature gradients and strains. For small temperature deviations from a fixed temperature T_o at the natural state, we may write

$$\theta = T_0 + T \quad , \quad |T| << T_0 \, , \; T_0 > 0 \,. \qquad (3.5.13)$$

In this case, retaining only up to quadratic terms in (3.5.3), we have

$$\Sigma_0 = \sigma_0 = S_0 - \rho_0 \eta_0 T - \frac{\rho_0 \gamma_0}{2 T_0} T^2$$

$$(3.5.14)$$

$$\sigma_1^{kl} = s_1^{kl} - b_1^{kl} T \quad , \qquad \sigma_3^{kl} = s_3^{kl} - b_3^{kl} T \,.$$

All other terms in (3.5.3) are not affected. Thus (3.5.8) and (3.5.9) become

$$t^{kl} = (1 - I_{\tilde{\varepsilon}}) s_1^{kl} - b_1^{kl} T + s_1^{ml} (\tilde{\varepsilon}_m{}^k - \varphi_m{}^k) \qquad (3.5.15)$$

$$+ s_1^{km} \varphi^l{}_m + \sigma_2^{klmn} \tilde{\varepsilon}_{mn} + \sigma_5^{klmn} \tilde{\gamma}_{mn}$$

$$
\begin{aligned}
m^{kl} = {} & (1 - I_{\tilde{\varepsilon}}) s_3^{kl} - b_3^{lk} T + s_3^{lm} (\tilde{\varepsilon}_m{}^k - \varphi_m{}^k) \\
& + s_3^{mk} \varphi^l{}_m + \sigma_4^{lkmn} \tilde{\gamma}_{mn} + \sigma_5^{mnlk} \tilde{\varepsilon}_{mn}
\end{aligned}
\tag{3.5.16}
$$

while (3.5.12) is unchanged. Now all constitutive coefficients are functions of $\tilde{x}$ for inhomogeneous solids and constant for the homogeneous solids.

The constitutive equations of linear micropolar thermoelasticity (3.5.12), (3.5.15) and (3.5.16) are further simplified when the material possesses various symmetries in its natural state. We give the result only for isotropic solids (first, without a center of symmetry). In this case, reduction is made by recalling the second, third, and fourth order isotropic tensors, e.g.,

$$
s_1^{kl} = s_1 g^{kl}
$$

$$
\sigma_2^{klmn} = \sigma_{21} g^{kl} g^{mn} + \sigma_{22} g^{km} g^{ln} + \sigma_{23} g^{kn} g^{lm}, \text{ etc.}
$$

Substituting these into (3.5.10), (3.5.15), and (3.5.16) gives

$$
\begin{aligned}
t^{kl} = {} & (1 - I_{\tilde{\varepsilon}}) s_1 g^{kl} - b_1 T g^{kl} + s_1 \tilde{\varepsilon}^{lk} + \sigma_{21} \tilde{\varepsilon}^r{}_r g^{kl} + \sigma_{22} \tilde{\varepsilon}^{kl} + \sigma_{23} \tilde{\varepsilon}^{lk} \\
& + \sigma_{51} \tilde{\gamma}^r{}_r g^{kl} + \sigma_{52} \tilde{\gamma}^{kl} + \sigma_{53} \tilde{\gamma}^{lk}
\end{aligned}
$$

$$m^{k\ell} = [(1 - I_{\tilde{\varepsilon}}) s_3 \, g^{k\ell} - b_3 T g^{k\ell} + s_3 \tilde{\varepsilon}^{\ell k} + \sigma_{41} \tilde{\gamma}^r{}_r g^{k\ell} + \tilde{\sigma}_{42} \gamma^{\ell k} + \tilde{\sigma}_{43} \gamma^{k\ell}$$

$$+ \sigma_{51} \tilde{\varepsilon}^r{}_r g^{k\ell} + \sigma_{52} \tilde{\varepsilon}^{\ell k} + \sigma_{53} \tilde{\varepsilon}^{\ell k}$$

$$q^k = \varkappa_1 \theta^k .$$

If the material is isotropic and possesses a center of symmetry (i.e., reflection is included in the material symmetry), then (noting that $\tilde{\underset{\sim}{\varepsilon}}$ is an absolute tensor and $\tilde{\underset{\sim}{\gamma}}$ is an axial tensor)

$$\sigma_{51} = \sigma_{52} = \sigma_{53} = s_3 = b_3 = 0 .$$

Thus the above equations reduce to

$$t^{k\ell} = (1 - I_{\tilde{\varepsilon}}) s_1 \, g^{k\ell} - b_1 T g^{k\ell} + \sigma_{21} \tilde{\varepsilon}^r{}_r g^{k\ell} + \sigma_{22} \tilde{\varepsilon}^{k\ell} + (\sigma_{23} + s_1) \tilde{\varepsilon}^{\ell k}$$

$$m^{k\ell} = \sigma_{41} \tilde{\gamma}^r{}_r g^{k\ell} + \tilde{\sigma}_{42} \gamma^{\ell k} + \sigma_{43} \tilde{\gamma}^{k\ell} . \qquad (3.5.17)$$

We note that a completely isotropic linear micropolar thermoelastic body cannot support an initial couple stress (cf. $(3.5.17)_2$. If the initial stress vanishes, then $S_1 = 0$, Writing

$$\sigma_{21} \equiv \lambda , \quad \sigma_{22} \equiv \mu + \varkappa , \sigma_{23} \equiv \mu , \; b_1 \equiv \beta_0$$

$$\sigma_{41} \equiv \alpha , \quad \alpha_{42} \equiv \gamma \quad , \quad \sigma_{43} \equiv \beta , \; \varkappa_1 \equiv \varkappa_1$$

the constitutive equations for isotropic solids (with a center of symmetry) become

$$
(3.5.18)\qquad
\begin{aligned}
t^{kl} &= -\beta_0 T\, g^{kl} + \lambda\, \tilde{\varepsilon}^{r}{}_{r}\, g^{kl} + (\mu + \varkappa)\, \tilde{\varepsilon}^{kl} + \mu\, \tilde{\varepsilon}^{lk} \\
m^{kl} &= \alpha\, \tilde{\gamma}^{r}{}_{r}\, g^{kl} + \beta\, \tilde{\gamma}^{kl} + \gamma\, \tilde{\gamma}^{lk} \\
q_k &= \varkappa_1\, \theta_{,k} \ .
\end{aligned}
$$

The free energy ψ now reads

$$
(3.5.19)\qquad
\begin{aligned}
\rho_0 \psi = {} & S_0 - \rho_0 \eta_0 T - \frac{\rho_0 \gamma_0}{2 T_0} T^2 - \beta_0 T \tilde{\varepsilon}^{r}{}_{r} + \frac{1}{2}\Big[\lambda \tilde{\varepsilon}^{k}{}_{k} \tilde{\varepsilon}^{l}{}_{l} + (\mu + \varkappa) \tilde{\varepsilon}^{k}{}_{l} \tilde{\varepsilon}_{k}{}^{l} + \\
& + \mu \tilde{\varepsilon}^{k}{}_{l} \tilde{\varepsilon}^{l}{}_{k}\Big] + \frac{1}{2}\left(\alpha \tilde{\gamma}^{k}{}_{k} \tilde{\gamma}^{l}{}_{l} + \beta \tilde{\gamma}^{k}{}_{l} \tilde{\gamma}^{l}{}_{k} + \gamma \tilde{\gamma}^{k}{}_{l} \tilde{\gamma}_{k}{}^{l}\right) .
\end{aligned}
$$

The constitutive coefficients are functions of θ only for the homogeneous materials.

Of course, the above results, $(3.5.18)_{1,2}$ and (3.5.19), could have been obtained more directly by linearizing (3.4.6), (3.4.7) and $(3.4.1)_3$. We pursued the present method to obtain the additional results (3.5.15), (3.5.16), and (3.5.12) for anisotropic solids. In fact, it is clear that (3.5.19) can be expressed entirely in terms of the invariants of $\tilde{\underset{\sim}{\varepsilon}}$ and $\tilde{\underset{\sim}{\gamma}}$, i.e.,

$$
(3.5.20)\qquad
\begin{aligned}
\rho_0 \psi = {} & S_0 - \rho_0 \eta_0 T - \frac{\rho_0 \gamma_0}{2 T_0} T^2 - \beta_0 T \operatorname{tr} \tilde{\underset{\sim}{\varepsilon}} + \frac{1}{2}\big[\lambda (\operatorname{tr} \tilde{\underset{\sim}{\varepsilon}})^2 \\
& + (\mu + \varkappa) \operatorname{tr} \tilde{\underset{\sim}{\varepsilon}} \tilde{\underset{\sim}{\varepsilon}}^{T} + \mu \operatorname{tr} \tilde{\underset{\sim}{\varepsilon}}^{2}\big] + \frac{1}{2}\big[\alpha (\operatorname{tr} \tilde{\underset{\sim}{\gamma}})^2 + \beta \operatorname{tr} \tilde{\underset{\sim}{\gamma}}^{2} \\
& + \gamma \operatorname{tr} \tilde{\underset{\sim}{\gamma}} \tilde{\underset{\sim}{\gamma}}^{T}\big]
\end{aligned}
$$

a result which also follows from writing the free energy in terms of the list of invariants (3.4.3) involving terms up to and including second order terms in $\tilde{\underset{\sim}{\varepsilon}}$ and $\tilde{\underset{\sim}{\gamma}}$.

The entropy η is given by

$$\eta = -\frac{\partial\Psi}{\partial T} = \eta_0 + \frac{\gamma_0}{T_0} T + \frac{\beta_0}{\rho_0} \operatorname{tr} \tilde{\underset{\sim}{\varepsilon}} \; . \tag{3.5.21}$$

<u>Incompressible materials.</u> In this case $\operatorname{tr} \tilde{\underset{\sim}{\varepsilon}} = 0$ and, according to (3.3.13), the stress constitutive equation is modified by a pressure term so that

$$t^{kl} = -(\beta_0 T + p) g^{kl} + (\mu + \varkappa) \tilde{\varepsilon}^{kl} + \varkappa \tilde{\varepsilon}^{kl}$$

$$\rho_0 \Psi = S_0 - \rho_0 \eta_0 T - \frac{\rho_0 \gamma_0}{2 T_0} T^2 + \frac{1}{2} [(\mu + \varkappa) \operatorname{tr} \tilde{\underset{\sim}{\varepsilon}} \tilde{\underset{\sim}{\varepsilon}}^T + \mu \operatorname{tr} \tilde{\underset{\sim}{\varepsilon}}^2]$$

$$+ \frac{1}{2} [\alpha (\operatorname{tr} \underset{\sim}{\gamma})^2 + \beta \operatorname{tr} \underset{\sim}{\gamma}^2 + \gamma \operatorname{tr} \underset{\sim}{\gamma} \underset{\sim}{\gamma}^T] \tag{3.5.22}$$

$$\eta = \eta_0 + \frac{\gamma_0}{T_0} T \; .$$

These equations are supplemented with the incompressibility condition

$$\operatorname{tr} \tilde{\underset{\sim}{\varepsilon}} = 0 \; . \tag{3.5.23}$$

Of course, the unknown pressure $p(x,t)$ will have to be determined upon integration of the field equations and use of the boundary conditions.

3.6. Nonnegative "Strain" Energy

For the isotropic linear theory (with vanishing initial stress), the internal energy $\varepsilon = \psi + \theta\,\eta$ may be written from (3.5.22) as

$$\rho_0 \varepsilon = S_0 + \rho_0 T_0 \eta_0 + \rho_0 \gamma_0 \left(T + \frac{T^2}{2T_0}\right) + \beta_0 T_0 \tilde{\varepsilon}^m{}_m + \rho_0 W \tag{3.6.1}$$

where

$$\begin{aligned} \rho_0 W \equiv {} & \frac{1}{2}\left[\lambda \left(\operatorname{tr} \tilde{\underset{\sim}{\varepsilon}}\right)^2 + (\mu + \varkappa) \operatorname{tr} \underset{\sim}{\tilde{\varepsilon}}\, \underset{\sim}{\tilde{\varepsilon}}^T + \mu \operatorname{tr} \underset{\sim}{\tilde{\varepsilon}}^2\right] \\ & + \frac{1}{2}\left[\alpha \left(\operatorname{tr} \underset{\sim}{\tilde{\gamma}}\right)^2 + \beta \operatorname{tr} \underset{\sim}{\tilde{\gamma}}^2 + \gamma \operatorname{tr} \underset{\sim}{\tilde{\gamma}}\, \underset{\sim}{\tilde{\gamma}}^T\right] . \end{aligned} \tag{3.6.2}$$

The scalar W is the internal energy when thermal effects are ignored, and, as such, we shall call W the "strain" energy for linear micropolar thermoelastic solids. For the linear anisotropic case, (3.6.1) and (3.6.2) become

$$\rho_0 \varepsilon = S_0 + \rho_0 T_0 \eta_0 + \rho_0 \gamma_0 \left(T + \frac{T^2}{2T_0}\right) + T_0 b_1^{kl} \tilde{\varepsilon}_{kl} + T_0 b_3^{kl} \tilde{\gamma}_{kl} + \rho_0 W \tag{3.6.3}$$

where

$$\rho_0 W \equiv \frac{1}{2} \sigma_2^{klmn} \tilde{\varepsilon}_{kl} \tilde{\varepsilon}_{mn} + \frac{1}{2} \sigma_4^{klmn} \tilde{\gamma}_{kl} \tilde{\gamma}_{mn} + \sigma_5^{klmn} \tilde{\varepsilon}_{kl} \tilde{\gamma}_{mn} . \tag{3.6.4}$$

A linear micropolar thermoelastic solid is said to be stable if

$$W \geq 0$$

for all possible states of strain. Notice that $W \geq 0$ is a weaker condition than $\varepsilon \geq 0$.

The imposition of stability for (3.6.2) has been studied by Eringen [1966 b] ; the results can be reproduced by noting that $\rho_0 W$ can be written as

$$\begin{aligned} 2\rho_0 W &= \tfrac{1}{3}(2\mu+3\lambda+\varkappa)\tilde{\varepsilon}^{k}_{\ k}\tilde{\varepsilon}^{\ell}_{\ \ell} + \varkappa\,\tilde{\varepsilon}^{[k\ell]}\tilde{\varepsilon}_{[k\ell]} \\ &+ (2\mu+\varkappa)[\tilde{\varepsilon}_{(k\ell)} - \tfrac{1}{3}\tilde{\varepsilon}^{m}_{\ m}g_{k\ell}][\tilde{\varepsilon}^{(k\ell)} - \tfrac{1}{3}\tilde{\varepsilon}^{n}_{\ n}g^{k\ell}] \\ &+ \tfrac{1}{3}(3\alpha+\beta+\gamma)\tilde{\gamma}^{k}_{\ k}\gamma^{\ell}_{\ \ell} + (\gamma-\beta)\tilde{\gamma}^{[k\ell]}\tilde{\gamma}_{[k\ell]} \\ &+ (\gamma+\beta)[\tilde{\gamma}_{(k\ell)} - \tfrac{1}{3}\tilde{\gamma}^{m}_{\ m}g_{k\ell}][\tilde{\gamma}^{(k\ell)} - \tfrac{1}{3}\tilde{\gamma}^{n}_{\ n}g^{k\ell}]. \end{aligned} \tag{3.6.5}$$

Since each term is a coefficient multiplied by a nonnegative factor involving the strains only, and since these factors can be varied independently of one another, we have proved :

Theorem : The necessary and sufficient conditions that (3.6.2) be nonnegative are

$$\begin{gathered} 0 \leq 2\mu+3\lambda+\varkappa,\ 0 \leq 2\mu+\varkappa,\ 0 \leq \varkappa \\ 0 \leq 3\alpha+\beta+\gamma,\ 0 \leq \gamma+\beta,\ 0 \leq \gamma-\beta. \end{gathered}$$

We shall henceforth demand that even for the anisotropic case, $\rho_0 W$ given by (3.6.4.) be nonnegative.

Chapter 4
FORMULATION OF PROBLEMS OF MICROPOLAR ELASTICITY

4.1. Scope

In this chapter first we collect all basic equations for the nonlinear theory, Art. 4.2, and those for the linear theory, (Art. 4.3), of micropolar thermoelasticity. The jump conditions , initial conditions and the pertinent kinematical formulae are listed. The field equations for the displacement, microrotation and temperature fields are obtained in Art. 4.4. In Art. 4.5 we present the proof of the uniqueness theorem for the linear case under rather a general set of boundary and initial conditions.

4.2. Résumé of Basic Equations of the Nonlinear Theory

The nonlinear theory of micropolar thermoelastic solids is based on the following equations:

Equations of balance (valid in $V-\sigma$)

$$\frac{\partial \rho}{\partial t} + (\rho v^k)_{;k} = 0 \tag{4.2.1}$$

$$\frac{\partial i^{kl}}{\partial t} + i^{kl}{}_{;m} v^m - i^{kr} \nu^l{}_{r} - i^{rl} \nu^k{}_{r} = 0 \tag{4.2.2}$$

(4.2.3) $$t^{k\ell}{}_{;k} + \rho\,(f^\ell - \dot{v}^\ell) = 0$$

(4.2.4) $$m^{k\ell}{}_{;k} + \varepsilon^{\ell mn}\,t_{mn} + \rho\,(\ell^\ell - \dot{\sigma}^\ell) = 0$$

(4.2.5) $$\rho\theta\dot{\eta} = q^k{}_{;k} + \rho h$$

(4.2.6) $$q^k\,\theta_{,k} \geq 0 \ .$$

Equations of jump (valid on σ) : On a discontinuity surface sweeping the body with a velocity $\underset{\sim}{u}$ we must have

(4.2.7) $$\underset{\sim}{[}\rho\,(\underset{\sim}{v} - \underset{\sim}{u})\underset{\sim}{]}\cdot \underset{\sim}{n} = 0$$

(4.2.8) $$\underset{\sim}{[}i^{k\ell}\chi_k{}^K\chi_\ell{}^L(\underset{\sim}{v} - \underset{\sim}{u})\underset{\sim}{]}\cdot \underset{\sim}{n} = 0$$

(4.2.9) $$\underset{\sim}{[}t^{k\ell} - \rho v^\ell(v^k - u^k)\underset{\sim}{]}\ n_k = 0$$

(4.2.10) $$\underset{\sim}{[}m^{k\ell} - \rho\,\sigma^\ell(v^k - u^k)\underset{\sim}{]}\,n_k = 0$$

(4.2.11) $$\underset{\sim}{[}t^{k\ell}v_\ell + m^{k\ell}\nu_\ell + q^k - (\rho\varepsilon + \tfrac{1}{2}\,\rho\,\underset{\sim}{v}\cdot\underset{\sim}{v} + \tfrac{1}{2}\rho\,\underset{\sim}{\sigma}\cdot\underset{\sim}{\nu})(v^k - u^k)\underset{\sim}{]}\,n_k = 0$$

(4.2.12) $$\underset{\sim}{[}\rho\,\eta(\underset{\sim}{v} - \underset{\sim}{u}) - \frac{\underset{\sim}{q}}{\theta}\underset{\sim}{]}\cdot \underset{\sim}{n} \geq 0 \ .$$

If σ coincides with the surface S of the body, then the quantities on the side S^+ are the prescribed values of loads, heat, and displacements. For example, in this case, for $[\underset{\sim}{t}^{kl}]\underset{\sim}{n}^{k}$ we write

$$t^{kl} n_k \equiv \overset{+}{t}^{kl} n_k \equiv t^{l}_{(n)}$$

where $t^{k}_{(n)}$ is the prescribed surface traction. This process leads to the boundary conditions on S.

<u>Constitutive equations</u> (anisotropic compressible solids) :

$$t^{kl} = \rho \frac{\partial \psi}{\partial \mathscr{E}_{KL}} x^{k}_{,K} \chi^{l}_{L} \tag{4.2.13}$$

$$m^{kl} = \rho \frac{\partial \psi}{\partial \Gamma_{LK}} x^{k}_{,K} \chi^{l}_{L} \tag{4.2.14}$$

$$q^{k} = Q^{K}(\underset{\sim}{\mathscr{E}}, \underset{\sim}{\Gamma}, \theta, \underset{\sim}{\nabla}\theta, \underset{\sim}{X})\, x^{k}_{,K} \tag{4.2.15}$$

$$\psi = \psi(\underset{\sim}{\mathscr{E}}, \underset{\sim}{\Gamma}, \theta, \underset{\sim}{X}) \tag{4.2.16}$$

$$\eta = -\frac{\partial \psi}{\partial \theta} \, . \tag{4.2.17}$$

<u>Constitutive equations</u> (anisotropic incompressible solids) :

$$t^{kl} = -p g^{kl} + \rho \frac{\partial \psi}{\partial \mathscr{E}_{KL}} x^{k}_{,K} \chi^{l}_{L} \tag{4.2.18}$$

(4.2.19) $$m^{k\ell} = \rho \frac{\partial \psi}{\partial \Gamma_{LK}} x^k{}_{,K} \chi^\ell{}_L$$

(4.2.20) $$q^k = Q^K(\underset{\sim}{\mathcal{C}}, \underset{\sim}{\Gamma}, \theta, \underset{\sim}{\nabla}\theta, \underset{\sim}{X})\, x^k{}_{,K}$$

(4.2.21) $$\psi = \psi(\underset{\sim}{\mathcal{C}}, \underset{\sim}{\Gamma}, \theta, \underset{\sim}{X})$$

(4.2.22) $$\eta = -\frac{\partial \psi}{\partial \theta}$$

subject to the incompressibility condition

(4.2.23) $$\det \mathcal{C} = 1 .$$

<u>Constitutive equations</u> (isotropic solids) :

(4.2.24) $$\underset{\sim}{t} = \alpha_1 \underset{\sim}{\mathcal{C}} + \alpha_2 \underset{\sim}{\mathcal{C}}^2 + \alpha_3 \underset{\sim}{\mathcal{C}}^3 + \alpha_4 \underset{\sim}{\mathcal{C}}\underset{\sim}{\mathcal{C}}^T + \alpha_5(\underset{\sim}{\mathcal{C}}\underset{\sim}{\mathcal{C}}^T\underset{\sim}{\mathcal{C}}^T + \underset{\sim}{\mathcal{C}}^2\underset{\sim}{\mathcal{C}}^T + \underset{\sim}{\mathcal{C}}\underset{\sim}{\mathcal{C}}^T\underset{\sim}{\mathcal{C}})$$
$$+ \alpha_6(\underset{\sim}{\mathcal{C}}^2\underset{\sim}{\mathcal{C}}^T\underset{\sim}{\mathcal{C}}^T + \underset{\sim}{\mathcal{C}}\underset{\sim}{\mathcal{C}}^T\underset{\sim}{\mathcal{C}}^T\underset{\sim}{\mathcal{C}}) + \alpha_7 \underset{\sim}{\mathcal{C}}\underset{\sim}{\gamma} + \alpha_8 \underset{\sim}{\mathcal{C}}\underset{\sim}{\gamma}^2 + \alpha_9(\underset{\sim}{\mathcal{C}}^2\underset{\sim}{\gamma} + \underset{\sim}{\mathcal{C}}\underset{\sim}{\gamma}\underset{\sim}{\mathcal{C}})$$

(4.2.25) $$\underset{\sim}{m} = \alpha_7 \underset{\sim}{\mathcal{C}}^2 + \alpha_8(\underset{\sim}{\mathcal{C}}^2\underset{\sim}{\gamma} + \underset{\sim}{\mathcal{C}}\underset{\sim}{\gamma}\underset{\sim}{\mathcal{C}}) + \alpha_9 \underset{\sim}{\mathcal{C}}^3 + \alpha_{10}\underset{\sim}{\mathcal{C}} + \alpha_{11}\underset{\sim}{\mathcal{C}}\underset{\sim}{\gamma} + \alpha_{12}\underset{\sim}{\mathcal{C}}\underset{\sim}{\gamma}^2 + \alpha_{13}\underset{\sim}{\mathcal{C}}\underset{\sim}{\gamma}^T$$
$$+ \alpha_{14}(\underset{\sim}{\mathcal{C}}\underset{\sim}{\gamma}^T\underset{\sim}{\gamma}^T + \underset{\sim}{\mathcal{C}}\underset{\sim}{\gamma}\underset{\sim}{\gamma}^T + \underset{\sim}{\mathcal{C}}\underset{\sim}{\gamma}^T\underset{\sim}{\gamma}) + \alpha_{15}(\underset{\sim}{\mathcal{C}}\underset{\sim}{\gamma}\underset{\sim}{\gamma}^T\underset{\sim}{\gamma}^T + \underset{\sim}{\mathcal{C}}\underset{\sim}{\gamma}^T\underset{\sim}{\gamma}^T\underset{\sim}{\gamma})$$

(4.2.26) $$\underset{\sim}{q} = Q^K(\underset{\sim}{\mathcal{C}}, \underset{\sim}{\Gamma}, \theta, \underset{\sim}{\nabla}\theta, \underset{\sim}{X})\, \underset{\sim}{x}_{,K}$$

(4.2.27) $$\psi = \psi(I_1, \ldots, I_{15}, \theta, \underset{\sim}{X}),$$ (see eq. (3.4.3) for I_α)

(4.2.28) $$\eta = -\frac{\partial \psi}{\partial \theta} .$$

For the incompressible solids, to the right-hand side of (4.2.24) one must add $-p\underset{\sim}{I}$ and set det $\underset{\sim}{d} = 1$ everywhere.

Kinematical relations :

$$\dot{v}^{\ell} = \frac{\partial v^{\ell}}{\partial t} + v^{\ell}{}_{;r} v^{r} \tag{4.2.29}$$

$$\nu^{k} = \Lambda_{k\ell}\, \dot{\varphi}^{\ell}\ , \ (\Lambda_{k\ell} \text{ given by } (1.6.19)) \tag{4.2.30}$$

$$\frac{D}{Dt} \chi^{k}{}_{K} = \nu^{k}{}_{\ell} \chi^{\ell}{}_{K} \tag{4.2.31}$$

$$\nu_{k\ell} = -\varepsilon_{k\ell m} \nu^{m} \tag{4.2.32}$$

$$\sigma^{k} = \dot{j}^{k\ell} \nu_{\ell} = (\dot{i}^{r}{}_{r} g^{k\ell} - \dot{i}^{k\ell}) \nu_{\ell} \tag{4.2.33}$$

$$\mathscr{C}_{KL} = x^{k}{}_{,K} \chi_{kL} \tag{4.2.34}$$

$$\Gamma_{KL} = \frac{1}{2} \varepsilon_{KMN} \chi^{\ell M}{}_{;L} \chi_{\ell}{}^{N} \tag{4.2.35}$$

$$\underset{\sim}{u} = \underset{\sim}{p} - \underset{\sim}{P}\ , \ \chi^{k}{}_{\ell} = \cos\sqrt{\varphi^{r}\varphi_{r}}\ \delta^{k}{}_{\ell} - \frac{\sin\sqrt{\varphi^{r}\varphi_{r}}}{\sqrt{\varphi^{r}\varphi_{r}}} \varepsilon^{k}{}_{\ell m} \varphi^{m} + \frac{1-\cos\sqrt{\varphi^{r}\varphi_{r}}}{\varphi^{r}\varphi_{r}} \varphi^{k} \varphi_{\ell} \tag{4.2.36}$$

$$Q^{K} \theta_{,K} \geq 0 \ \text{(thermodynamical)} \tag{4.2.37}$$

$$W \geq 0 \ \text{(stability)}\,. \tag{4.2.38}$$

Both inequalities must be satisfied for all independent processes.

Boundary conditions (valid on the surface $S-\sigma$ of the body :

(4.2.39) $$t^{kl} n_k = \bar{t}^{\,l} \quad \text{on} \quad S_T$$

(4.2.40) $$m^{kl} n_k = \bar{m}^{\,l} \quad \text{on} \quad S_T$$

(4.2.41) $$u^k = \bar{u}^k \quad \text{on} \quad S_u = S - S_T$$

(4.2.42) $$\varphi^k = \bar{\varphi}^k \quad \text{on} \quad S_u$$

(4.2.43) $$\underset{\sim}{q} \cdot \underset{\sim}{n} = q_{(\underset{\sim}{n})} \quad \text{on} \quad S_q$$

(4.2.44) $$\theta = \bar{\theta} \quad \text{on} \quad S - S_q$$

where the quantities carrying a superposed bar are prescribed on a part of the body. The above mixed boundary conditions represent only the conditions for a possible class of problems ; clearly other possibilities exist. A full discussion of all cases requires the proof of existence and uniqueness theorems. For example, in heat conduction problems, another important condition replacing one of (4.2.43) and (4.2.44) is the radiation from a part S_r of the boundary to the outside

(4.2.45) $$\underset{\sim}{q} \cdot \underset{\sim}{n} + r(\theta - \theta_1) = 0$$

where r is a function of the difference between the temperatures near the surface of the body and outside.

Initial conditions (valid in $V - \sigma$) :

$$\underset{\sim}{u}(\underset{\sim}{x}, 0) = \underset{\sim}{u}^{\circ}(\underset{\sim}{x}) \tag{4.2.46}$$

$$\underset{\sim}{v}(\underset{\sim}{x}, 0) = \underset{\sim}{v}^{\circ}(x) \tag{4.2.47}$$

$$\underset{\sim}{\varphi}(\underset{\sim}{x}, 0) = \underset{\sim}{\varphi}^{\circ}(x) \tag{4.2.48}$$

$$\underset{\sim}{\nu}(\underset{\sim}{x}, 0) = \underset{\sim}{\nu}^{\circ}(\underset{\sim}{x}) \tag{4.2.49}$$

$$\underset{\sim}{\theta}(\underset{\sim}{x}, 0) = \overset{\circ}{\underset{\sim}{\theta}}(\underset{\sim}{x}) \tag{4.2.50}$$

where the quantities carrying a superscript (°) are prescribed throughout the body $V - \sigma$.

4.3. Résumé of Basic Equations of Linear Micropolar Thermoelasticity

Equations of balance : Same as (4.2.1) to (4.2.6).

Equations of Jump : Same as (4.2.7) to (4.2.12).

Constitutive equations (anisotropic compressible solids ; vanishing initial stresses) :

$$t^{kl} = -b_1^{kl}\, T + \sigma_2^{klmn\sim}\, \varepsilon_{mn} + \sigma_5^{klmn\sim}\, \gamma_{mn} \tag{4.3.1}$$

$$m^{kl} = -b_3^{lk}\, T + \sigma_4^{lkmn\sim}\, \gamma_{mn} + \sigma_5^{mnlk\sim}\, \varepsilon_{mn} \tag{4.3.2}$$

(4.3.3) $$q^k = \varkappa_1^{kl} T_{,l}$$

(4.3.4) $$\rho_0 \psi = S_0 - \rho_0 \eta_0 T - \frac{\rho_0 \gamma_0}{2T_0} T^2 - T b_1^{kl} \tilde{\varepsilon}_{kl} + \frac{1}{2} \sigma_2^{klmn} \tilde{\varepsilon}_{kl} \tilde{\varepsilon}_{mn}$$
$$- T b_3^{kl} \tilde{\gamma}_{kl} + \frac{1}{2} \sigma_4^{klmn} \tilde{\gamma}_{kl} \tilde{\gamma}_{mn} + \sigma_5^{klmn} \tilde{\varepsilon}_{kl} \tilde{\gamma}_{mn}$$

(4.3.5) $$\eta = \eta_0 + \frac{\gamma_0}{T_0} T + \frac{1}{\rho_0} b_1^{kl} \tilde{\varepsilon}_{kl} + \frac{1}{\rho_0} b_3^{kl} \tilde{\gamma}_{kl} \quad .$$

Constitutive equations for incompressible solids follow from these by taking $I_{\tilde{\varepsilon}} = 0$ and adding $-p\, g^{kl}$ on the right-hand side of (4.3.1).

Constitutive equations (isotropic compressible solids) :

(4.3.6) $$t^{kl} = (-\beta_0 T + \lambda \tilde{\varepsilon}^r{}_r) g^{kl} + (\mu + \varkappa) \tilde{\varepsilon}^{kl} + \mu \tilde{\varepsilon}^{lk}$$

(4.3.7) $$m^{kl} = \alpha \tilde{\gamma}^r{}_r g^{kl} + \beta \tilde{\gamma}^{kl} + \gamma \tilde{\gamma}^{lk}$$

(4.3.8) $$q^k = \varkappa_1 T_{,k}$$

(4.3.9) $$\rho_0 \psi = S_0 - \rho_0 \eta_0 T - \frac{\rho_0 \gamma_0}{2T_0} T^2 - \beta_0 T \tilde{\varepsilon}^r{}_r + \frac{1}{2} [\lambda \tilde{\varepsilon}^k{}_k \tilde{\varepsilon}^l{}_l + (\mu + \varkappa) \tilde{\varepsilon}^k{}_l \tilde{\varepsilon}^l{}_k$$
$$+ \mu \tilde{\varepsilon}^k{}_l \tilde{\varepsilon}^l{}_k] + \frac{1}{2} (\alpha \gamma^k{}_k \gamma^l{}_l + \beta \gamma^k{}_l \gamma^l{}_k + \gamma \gamma^k{}_l \gamma_k^l)$$

$$\eta = \eta_0 + \frac{\gamma_0}{T_0} T + \frac{\beta_0}{\rho_0} \tilde{\varepsilon}^{r}_{\ r} \quad . \tag{4.3.10}$$

For incompressible solids, we set $\tilde{\varepsilon}^{r}_{\ r} = 0$ and add $-pg^{k\ell}$ to the right-hand side of (4.3.6).

Kinematical relations :

$$\dot{v}^{\ell} \simeq \frac{\partial v^{\ell}}{\partial t} \tag{4.3.11}$$

$$\nu^{k} \simeq \dot{\varphi}^{\ell} \tag{4.3.12}$$

$$\nu_{k\ell} \simeq -\varepsilon_{k\ell m} \dot{\varphi}^{m} \tag{4.3.13}$$

$$\sigma^{k} = j^{\cdot k\ell} \nu_{\ell} \simeq (i^{\cdot r}_{\ r} g^{k\ell} - i^{\cdot k\ell}) \dot{\varphi}_{\ell} \tag{4.3.14}$$

$$\varepsilon_{k\ell} \simeq u_{\ell;k} - \varepsilon_{k\ell m} \varphi^{m} \tag{4.3.15}$$

$$\gamma_{k\ell} \simeq \varphi_{k;\ell} \quad . \tag{4.3.16}$$

Restriction on constitutive moduli :

$$\begin{gathered} 3\lambda + 2\mu + \varkappa \geq 0 \ , \quad 2\mu + \varkappa \geq 0 \ , \quad \varkappa \geq 0 \\ 3\alpha + \beta + \gamma \geq 0 \ , \quad \gamma + \beta \geq 0 \ , \quad \gamma - \beta \geq 0 \ ; \varkappa_1 \geq 0 \ . \end{gathered} \tag{4.3.17}$$

Boundary conditions : Same as in (4.2.39) to (4.2.45).

Initial conditions : Same as in (4.2.46) to (4.2.50).

4.4. Field Equations of the Linear Theory

In the linear theory of micropolar elasticity we can write partial differential equations for the displacement and microrotation fields $\underset{\sim}{u}$ and $\underset{\sim}{\varphi}$ by substituting (4.3.1) and (4.3.2) into (4.2.3) and (4.2.4). Hence

$$(4.4.1) \qquad (\overset{klmn}{\underset{2}{\sigma}} \overset{\sim}{\varepsilon}_{mn} + \overset{klmn}{\underset{5}{\sigma}} \overset{\sim}{\gamma}_{mn} - \overset{kl}{\underset{1}{b}} T)_{;k} + \rho(f^{l} - \ddot{u}^{l}) = 0$$

$$(4.4.2) \qquad (\overset{lkmn}{\underset{4}{\sigma}} \overset{\sim}{\gamma}_{mn} + \overset{mnlk}{\underset{5}{\sigma}} \overset{\sim}{\varepsilon}_{mn} - \overset{lk}{\underset{2}{b}} T)_{;k} + \varepsilon^{lmn}(\overset{rs}{\underset{2}{\sigma}}_{mn} \overset{\sim}{\varepsilon}_{rs}$$

$$+ \overset{rs}{\underset{5}{\sigma}}_{mn} \gamma_{rs} - b_{1mn} T) + \rho(l^{l} - j^{lk} \ddot{\varphi}_{k}) = 0$$

where we also used (4.2.29) to (4.2.33) and linearized $\dot{\sigma}^{l}$ and $\dot{\upsilon}^{l}$. Note that in the linear theory both ρ and j^{kl} are constants to the lowest degree of approximation since they appear in products with $\underset{\sim}{\gamma}$ and $\underset{\sim}{\nu}$ respectively.

The heat conduction equation is obtained by substituting (4.3.5) into (4.2.5). Thus

$$(4.4.3) \qquad \rho_{0} \gamma_{0} \dot{T} + T_{0} \overset{kl}{\underset{1}{b}} \overset{\sim}{\dot{\varepsilon}}_{kl} + T_{0} \overset{kl}{\underset{3}{b}} \overset{\sim}{\dot{\gamma}}_{kl} - (\overset{kl}{\underset{1}{\varkappa}} T_{,l})_{;k} - \rho h = 0 .$$

For the time rate we have, in the linear theory,

$$T \equiv \frac{\partial \dot{T}}{\partial t} , \; etc.$$

The corresponding field equations for isotropic solids are :

$$(\lambda+\mu)u^{l}{}_{;l}{}^{k}+(\mu+\varkappa)u^{k}{}_{;l}{}^{l}+\varkappa\varepsilon^{klm}\varphi_{m;l}-\beta_0 T^{;k}+\rho(f^k-\ddot{u}^k)=0 \quad (4.4.4)$$

$$(\alpha+\beta)\varphi^{l}{}_{;l}{}^{k}+\gamma\varphi^{k}{}_{;l}{}^{l}+\varkappa\varepsilon^{klm}u_{m;l}-2\varkappa\varphi^k+\rho(l^k-j^{kl}\ddot{\varphi}_l)=0 \quad (4.4.5)$$

$$\rho_0\gamma_0\dot{T}+\beta_0 T_0\dot{u}^{r}{}_{;r}-(\varkappa T,^{k})_{;k}-\rho h=0\,. \quad (4.4.6)$$

Equations (4.4.1) to (4.4.3) are the partial differential equations satisfied by the fields u^k, φ^k, and T. Equations (4.4.1) and (4.4.2) with $\beta_0\equiv 0$ were given by Eringen and Suhubi [1964] and recapitulated with supplementary theorems (thermodynamical and uniqueness, etc.) by Eringen [1966 b]. The thermoelastic theory was pursued by Nowacki, Tauchert, et al. [1968], and others. Professor Nowacki in a series of papers [1969], extended many results of the classical theory of thermoelasticity. The literature on the linear theory of micropolar elasticity is extensive. The number of papers exceeds several hundred, and it is impossible to do justice to all contributors by a brief mention of their contributions here. Neither time nor space permits this. This field, since its birth some six or so years ago, has become a truly international field. Under one cloak or the other it has found many serious followers in America, Europe, and Asia, particularly in the U.S.A., England, Ireland, Germany, Italy, Poland, Yugoslavia, Rumania, Turkey, USSR, India, and

Japan.

For some purposes, the vectorial form of (4.4.4) to (4.4.6) may be more useful. These are obtained by recalling the operators :

$$u^{\ell}{}_{;\ell}{}^{k}\,\underset{\sim}{g}_k = \underset{\sim}{\nabla}\underset{\sim}{\nabla}\cdot\underset{\sim}{u} \quad , \quad \varepsilon^{k\ell m}\varphi_{m;\ell}\,\underset{\sim}{g}_k = \underset{\sim}{\nabla}\times\underset{\sim}{\varphi}$$

$$u^{k}{}_{;\ell}{}^{\ell}\,\underset{\sim}{g}_k = \underset{\sim}{\nabla}\underset{\sim}{\nabla}\cdot\underset{\sim}{u} - \nabla\times\underset{\sim}{\nabla}\times\underset{\sim}{u} \quad , \quad \underset{\sim}{\nabla} \equiv g^k \frac{\partial}{\partial x^k} \quad .$$

Thus

$$(4.4.7)\ (\lambda+2\mu+\varkappa)\underset{\sim}{\nabla}\underset{\sim}{\nabla}\cdot\underset{\sim}{u} - (\mu+\varkappa)\underset{\sim}{\nabla}\times\underset{\sim}{\nabla}\times\underset{\sim}{u} + \varkappa\underset{\sim}{\nabla}\times\underset{\sim}{\varphi} - \beta_0\underset{\sim}{\nabla}T + \rho(\underset{\sim}{f} - \ddot{\underset{\sim}{u}}) = \underset{\sim}{0}$$

$$(4.4.8)\ (\alpha+\beta+\gamma)\underset{\sim}{\nabla}\underset{\sim}{\nabla}\cdot\underset{\sim}{\varphi} - \gamma\underset{\sim}{\nabla}\times\underset{\sim}{\nabla}\times\underset{\sim}{\varphi} + \varkappa\underset{\sim}{\nabla}\times\underset{\sim}{u} - 2\varkappa\underset{\sim}{\varphi} + \rho(\underset{\sim}{\ell} - \dot{\underset{\approx}{j}}\cdot\ddot{\underset{\sim}{\varphi}}) = \underset{\sim}{0}$$

$$(4.4.9)\qquad \rho_0\gamma_0\dot{T} + \beta_0 T_0\underset{\sim}{\nabla}\cdot\dot{\underset{\sim}{u}} - \underset{\sim}{\nabla}\cdot(\varkappa\underset{\sim}{\nabla}T) - \rho h = 0 \quad .$$

These equations under certain boundary and initial conditions can be shown to possess unique solutions under the values of the thermoelastic moduli restricted by (4.3.17). As the initial conditions, we consider

$$(4.4.10)\qquad \begin{aligned} \underset{\sim}{u}(\underset{\sim}{x},0) &= \underset{\sim}{u}^{\circ}(\underset{\sim}{x}) \\ \dot{\underset{\sim}{u}}(\underset{\sim}{x},0) &= \underset{\sim}{v}^{\circ}(\underset{\sim}{x}) \\ \underset{\sim}{\varphi}(\underset{\sim}{x},0) &= \underset{\sim}{\varphi}^{\circ}(\underset{\sim}{x}) \\ \dot{\underset{\sim}{\varphi}}(\underset{\sim}{x},0) &= \underset{\sim}{\nu}^{\circ}(\underset{\sim}{x}) \\ T(\underset{\sim}{x},0) &= T^{\circ}(\underset{\sim}{x}) \end{aligned} \qquad \text{in } V-\sigma$$

where $\underset{\sim}{u}^{\circ}, \underset{\sim}{v}^{\circ}, \underset{\sim}{\varphi}^{\circ}$ and $\underset{\sim}{\nu}^{\circ}$ and T° are prescribed in $V-\sigma$. The boundary conditions take a multitude of different forms. One such set is given by (4.2.39) to (4.2.45). In fact, it can be shown that (see section below) admissible boundary conditions allowing unique solutions are derivable by satisfying

$$\underset{\sim}{\Big[} \underset{\sim}{t}^{k}_{(\underset{\sim}{n})} u_k + m^{k}_{(\underset{\sim}{n})} \dot{\varphi}_k + \frac{q_{(n)} T}{T_0} \underset{\sim}{\Big]} = 0 \quad \text{on } S \tag{4.4.11}$$

where the boldface bracket indicates the jump of the quantity on the surface S of the body.

4.5. Uniqueness Theorem

Theorem 1, (uniqueness) : If the conditions (4.3.17), $r_k \neq \varphi_k$ are satisfied and $\dot{j}^{k\ell}$ is a positive definite matrix in a bounded, regular domain*) V of space with boundary S, then there exists at most one $\underset{\sim}{u}(\underset{\sim}{x}, t)$ one $\varphi(\underset{\sim}{x}, t)$ and one $T(\underset{\sim}{x}, t)$ all twice continuously differentiable for $\underset{\sim}{x}$ in $V+S$ and $0 \leq t \leq \infty$, which satisfy (4.4.1) to (4.4.3), the initial conditions (4.4.10), and a set of boundary conditions compatible with (4.4.11).

---

*) The terminology "regular domain of space" denotes the domains for which the Green-Gauss theorem is valid.

Proof : Suppose that the contrary is valid and two solutions $\underset{\sim}{u}^{(\alpha)}, \underset{\sim}{\varphi}^{(\alpha)}, T^{(\alpha)}$, $\alpha = 1,2$, exist satisfying (4.4.1) to (4.4.3) and (4.4.10) to (4.4.11). Let

(4.5.1) $$\underset{\sim}{u} \equiv \underset{\sim}{u}^{(1)} - \underset{\sim}{u}^{(2)}, \quad \underset{\sim}{\varphi} \equiv \underset{\sim}{\varphi}^{(1)} - \underset{\sim}{\varphi}^{(2)}, \quad T \equiv T^{(1)} - T^{(2)}.$$

Then clearly $\underset{\sim}{u}, \underset{\sim}{\varphi}, T$ satisfy (4.4.1) to (4.4.3) and (4.4.10) to (4.4.11) with $\underset{\sim}{f} = \underset{\sim}{l} = \underset{\sim}{0}$, $h = 0$, $\underset{\sim}{u}^{\circ} = \underset{\sim}{v}^{\circ} = \underset{\sim}{\varphi}^{\circ} = \underset{\sim}{\nu}^{\circ} = 0$, $T^{\circ} = 0$. Further, from (3.6.3) and (3.6.4)

$$\rho_0 \dot{\varepsilon} = (\rho_0 \gamma_0 \dot{T} + T_0 b_1^{kl} \dot{\tilde{\varepsilon}}_{kl} + T_0 b_3^{kl} \dot{\tilde{\gamma}}_{kl}) + \frac{\rho_0 \gamma_0}{T_0} T \dot{T} + \rho_0 \dot{W} .$$

Using (4.4.3) and (4.3.3) for q^k in this, we have

(4.5.2) $$\rho_0 \dot{\varepsilon} = q^k{}_{;k} + \frac{\rho_0 \gamma_0}{T_0} T \dot{T} + \rho_0 \dot{W} .$$

Also, employing (4.3.1) and (4.3.2) in

$$\rho_0 \dot{W} = \sigma_2^{klmn} \dot{\tilde{\varepsilon}}_{kl} \tilde{\varepsilon}_{mn} + \sigma_4^{klmn} \dot{\tilde{\gamma}}_{kl} \tilde{\gamma}_{mn} + \sigma_5^{klmn} \dot{\tilde{\varepsilon}}_{kl} \tilde{\gamma}_{mn} + \sigma_5^{klmn} \tilde{\varepsilon}_{kl} \dot{\tilde{\gamma}}_{mn}$$

gives

$$\rho_0 \dot{W} = (t^{kl} + b_1^{kl} T) \dot{\tilde{\varepsilon}}_{kl} + (m^{lk} + b_3^{kl} T) \dot{\tilde{\gamma}}_{kl} .$$

Using this last expression in (4.5.2) yields (taking cognizance of (4.4.3) again)

$$\rho_0 \dot{\varepsilon} = \frac{\theta}{T_0} q^k_{;k} + t^{k\ell} \tilde{\dot{\varepsilon}}_{k\ell} + m^{\ell k} \tilde{\dot{\gamma}}_{k\ell} \ . \tag{4.5.3}$$

Subtracting (4.5.2) from (4.5.3) gives

$$\rho_0 \frac{d}{dt}\left(W + \frac{\gamma_0}{2T_0} T^2\right) = \frac{T}{T_0} q^k_{;k} + t^{k\ell}(\dot{u}_{\ell;k} + \nu_{k\ell}) + m^{\ell k} \dot{\varphi}_{k;\ell} \ .$$

Whence, from (4.4.1) and (4.4.2) we have

$$\rho_0 \frac{d}{dt}\left(W + \frac{1}{2} v^2 + \frac{1}{2} j_{k\ell} \nu^k \nu^\ell + \frac{\gamma_0}{2T_0} T^2\right) = \frac{T}{T_0} q^k_{;k} + (t^{k\ell} \dot{u}_\ell)_{;k} \ .$$
$$+ (m^{\ell k} \dot{\varphi}_k)_{;\ell}$$

Or finally (noting that for the linear theory $\rho \simeq \rho_0, j_{k\ell} \simeq g_k{}^K g_\ell{}^L J_{KL}$)

$$\frac{d}{dt} \int_V \rho \left(W + \frac{1}{2} v^2 + \frac{1}{2} j_{k\ell} \nu^k \nu^\ell + \frac{\gamma_0}{2T_0} T^2\right) dv = \tag{4.5.4}$$
$$- \frac{1}{T_0} \int_V q^k T_{;k} \, dv + \oint_S \left(t^{k\ell} \dot{u}_\ell + m^{k\ell} \dot{\varphi}_\ell + \frac{q^k T}{T_0}\right) n_k \, da \ .$$

From the boundary conditions, the surface integral in (4.5.4) vanishes ; also, from the Clausius-Duhem inequality, $q^k T_{;k} \geq 0$, whence

$$\frac{d}{dt} \int_V \rho \left(W + \frac{1}{2} v^2 + \frac{1}{2} j_{k\ell} \nu^k \nu^\ell + \frac{\gamma_0}{2T_0} T^2\right) dv \leq 0 \ .$$

But at $t = 0$ from the initial conditions, this last volume integral vanishes and thus

$$\int_V \rho\left(W + \frac{1}{2} v^2 + \frac{1}{2} j_{k\ell} \nu^k \nu^\ell + \frac{\gamma_0}{2T_0} T^2\right) dv \leq 0 .$$

Now W is nonnegative definite and $\frac{1}{2} v^2 + \frac{1}{2} j_{k\ell} \nu^k \nu^\ell + \frac{\gamma_0}{2T_0} T^2$

is positive definite, therefore, the only possibility for the satisfaction of this inequality is

$$v^k = \nu^k = 0, \; T = 0 .$$

Or finally

$$\begin{aligned} \underset{\sim}{u}^{(1)}(\underset{\sim}{x}, t) &= \underset{\sim}{u}^{(2)}(\underset{\sim}{x}, t) \\ \underset{\sim}{\varphi}^{(1)}(\underset{\sim}{x}, t) &= \underset{\sim}{\varphi}^{(2)}(\underset{\sim}{x}, t) \\ T^{(1)}(\underset{\sim}{x}, t) &= T^{(2)}(\underset{\sim}{x}, t) \end{aligned}$$

and thus the proof of the theorem.

For the general micromorphic theory, the uniqueness theorem was proved by Soos [1969].

REFERENCES

Coleman, B.D. and W. Noll [1961]: "Foundations of Linear Viscoelasticity," Rev. Modern Phys. 33, 239-249.

Cosserat, E. and F. Cosserat [1909] : Théorie des Corps Déformables. Paris A. Hermann.

Duhem, P. [1893] : "Le potential thermodynamique et la pression hydrostatique, Ann. Ecole Norm. 10, 187-230.

Eringen, A.C. [1962] : Nonlinear Theory of Continuous Media, Arts. 32,40. McGraw-Hill.

Eringen, A.C. and E.S. Suhubi [1964 a] : "Nonlinear Theory of Simple Microelastic Solids I & II", Int. J. Engng. Sci. 189-203 and 389-404.

Eringen, A.C. [1964 b] : "Simple Microfluids", Int. J. Engng. Sci. 2, 205-217.

Eringen, A.C. [1966 a] : "A Unified Theory of Thermomechanical Materials", Int. J. Engng. Sci. 4, 179-202.

Eringen, A.C. [1966 b] : "Linear Theory of Micropolar Elasticity", J. Math. & Mech. 15, 6, 909-924.

Eringen, A.C. [1967 a] : "Mechanics of Micromorphic Continua", Mechanics of Generalized Continua. 18-35. Edited by E. Kröner. Springer-Verlag,1968.

Eringen, A.C. [1967 b] : "Linear Theory of Micropolar Viscoelasticity", Int. J. Engng. Sci., 5, 2, 191-204.

Eringen, A.C. [1967 c] : "Compatibility Conditions of the Theory of Micromorphic Elastic Solids", NASA report (1967). See also J. Math. and Mech. 19, 6, 473-481, 1969.

Eringen, A.C. [1967 b] : Mechanics of Continua. New York. John Wiley & Sons.

Eringen, A.C. [1968] : Theory of Micropolar Elasticity Fracture vol. II, 622-729, edit. By Liebowitz, Academic Press.

Eringen, A.C. and J. Ingram : "A Continuum Theory of Chemically Reacting Media," Int. J. Engng. Sci., 3, 197-212 and 5, 289-322 (1967).

Kafadar, C.B. and A.C. Eringen [1970] : "Polar Media - The Classical and Relativistic Theory", Office of Naval Research Tech. Report.

MacCullagh, J. [1839] : "An essay Towards a Dynamical Theory of Crystalline Reflection and Refraction", Trans. Roy. Irish Acad. Sci., 21, 17-50.

Nowacki, W. and W.K. Nowacki [1969] : "Generation of Waves in Infinite Micropolar Elastic Solid Body I, II", Bull. Acad. Pol. Sci. Techn., 17, 39-47, 49-56.

Sandru, N. [1966] : "On Some Problems of the Linear Theory of Asymmetric Elasticity", Int. J. Engng. Sci., 4, 81-96.

Soos, E. [1969] : "Uniqueness Theorems for Homogeneous, Isotropic, Simple Elastic and Thermoelastic Materials Having Microstructure", Int. J. Engng. Sci., 7, 257-268.

Stajanowic, R. [1969] : "Mechanics of Polar Continua", lecture notes, Int. Center of Mechanical Sciences, 1969.

Tauchert, T.R., W.D. Claus, and T. Ariman [1968] : "The Linear Theory of Micropolar Thermoelasticity", Int. J. Engng. Sci., 6, 37-47.

Truesdell, C. and W. Noll [1965] : "The Nonlinear Field Theories of Mechanics", Handbuch der Physik, Bd. III/3, Springer-Verlag, Berlin.

Voigt, W. [1887] : "Theoritishe Studien uber die Elastizitatsverhaltnisse der Krystalle, Abh. Wiss. Ges. Gottingen, 34.

Voigt, W. [1895] : Kompendium der theoretischen Physik, Bd. 1, Leipzig.

TABLE OF CONTENTS

Page